Honda CB650 sohc Fours Owners Workshop Manual

by Martyn Meek
with an additional Chapter on the later models
by Mark Coombs

Models covered
CB650 Z. 626cc. UK November 1978 to February 1982
CB650 SC Custom/Nighthawk. 626cc. UK February 1982 to February 1984
CB650 Z. 626cc. US January 1979 to December 1979
CB650 A. 626cc. US January 1980 to December 1980
CB650 CA Custom. 626cc. US October 1979 to December 1980
CB650 B. 626cc. US January 1981 to December 1981
CB650 CB Custom. 626cc. US January 1981 to December 1981
CB650 C. 626cc. US January 1982 to December 1982
CB650 SC Nighthawk. 626cc. US October 1981 to December 1982

ISBN 978 1 85010 759 0

Disclaimer

There are risks associated with automotive repairs. The ability to make repairs depends on the individual's skill, experience and proper tools. Individuals should act with due care and acknowledge and assume the risk of performing automotive repairs.

The purpose of this manual is to provide comprehensive, useful and accessible automotive repair information, to help you get the best value from your vehicle. However, this manual is not a substitute for a professional certified technician or mechanic.

This repair manual is produced by a third party and is not associated with an individual vehicle manufacturer. If there is any doubt or discrepancy between this manual and the owner's manual or the factory service manual, please refer to the factory service manual or seek assistance from a professional certified technician or mechanic.

Even though we have prepared this manual with extreme care and every attempt is made to ensure that the information in this manual is correct, neither the publisher nor the author can accept responsibility for loss, damage or injury caused by any errors in, or omissions from, the information given.

THE BOOK ®

Haynes Group Limited
Haynes North America, Inc

www.haynes.com

British Library Cataloguing in Publication Data
Meek, Martyn   Honda CB650 sohc fours owners workshop manual. 1. Motorcycles. Maintenance I. Title   II. Series 629.28775 ISBN 1–85010–759–9
Library of Congress Catalog Card Number
90–86328

Acknowledgements

Our thanks are due to Paul Branson Motorcycles, of Yeovil, who loaned the machine featured in the photographs throughout this manual. Alan Jackson assisted with the stripdown and rebuilding of the machine, and devised many of the ingenious methods used to overcome the lack of service tools. Les Brazier arranged and took the photographs which accompany the text.

Finally, we would also like to thank the Avon Rubber Company who provided information on tyre fitting, Renold Ltd for information about replacement chains, and NGK Spark Plugs (UK) Ltd, who furnished advice about sparking plug conditions.

About this manual

The purpose of this manual is to present the owner with a concise and graphic guide which will enable him to tackle any operation from basic routine maintenance to a major overhaul. It has been assumed that any work would be undertaken without the luxury of a well-equipped workshop and a range of manufacturer's service tools.

To this end, the machine featured in the manual was stripped and rebuilt in our own workshop, by a team comprising a mechanic, a photographer and the author. The resulting photographic sequence depicts events as they took place, the hand shown being those of the author and the mechanic.

The use of specialised, and expensive, service tools was avoided unless their use was considered to be essential due to risk of breakage or injury. There is usually some way of improvising a method of removing a stubborn component, provided that a suitable degree of care is exercised.

The author learnt his motorcycle mechanics over a number of years, faced with the same difficulties and using similar facilities to those encountered by most owners. It is hoped that this practical experience can be passed on through the pages of this manual.

Where possible, a well-used example of the machine is chosen for the workshop project, as this highlights any areas which might be particularly prone to giving rise to problems. In this way, any such difficulties are encountered and resolved before the text is written, and the techniques used to deal with them can be incorporated in the relevant section. Armed with a working knowledge of the machine, the author undertakes a considerable amount of research in order that the maximum amount of data can be included in this manual.

Each Chapter is divided into numbered sections. Within these sections are numbered paragraphs. Cross reference throughout the manual is quite straightforward and logical. When reference is made 'See Section 6.10' means Section 6, paragraph 10 in the same Chapter. If another Chapter were intended the reference would read, for example, 'See Chapter 2, Section 6.10'. All the photographs are captioned with a section/paragraph number to which they refer and are relevant to the Chapter text adjacent.

Figures (usually line illustrations) appear in a logical but numerical order, within a given Chapter. Fig. 1.1 therefore refers to the first figure in Chapter 1.

Left-hand and right-hand descriptions of the machines and their components refer to the left and right of a given machine when the rider is seated normally.

Motorcycle manufacturers continually make changes to specifications and recommendations, and these, when notified, are incorporated into our manuals at the earliest opportunity.

We take great pride in the accuracy of information given in this manual, but motorcycle manufacturers make alterations and design changes during the production run of a particular motorcycle of which they do not inform us. No liability can be accepted by the authors or publishers for loss, damage or injury caused by any errors in, or omissions from, the information given.

Contents

Right-hand view of 1980 UK Honda CB650 Z

Left-hand view of 1980 UK Honda CB650 Z

Engine unit of 1980 UK Honda CB650 Z

Introduction to the Honda CB650 sohc Fours

The present Honda empire, which started in a wooden shack in 1947, now occupies vast factory space in its native Japan, covering all aspects of a modern motorcycle design, research, testing and production. In addition, Honda now have large factory space in the USA and on the Continent.

Although for many years manufacturers have produced what are virtually replicas of the 'Works' racing machines, for sale in the general public, it is true to say that no machine captured the interest of the motor cyclist so vividly as the first production version of the Honda 4 cylinder. Already a legend in racing circles, the Honda 4 had represented a serious challenge whenever it appeared in international events. With riders such as the late Bob MacIntyre, Jim Redman and Mike Hailwood, the 4 demonstrated its supremacy on frequent occasions, irrespective of whether the 125 cc, 250 cc, 350 cc or 500 cc version was raced. Even the previously unbeaten multi-cylinder Italian models no longer had things their own way and were hard put to continue racing under truly competitive terms.

When, at the end of 1967, Honda withdrew from racing and commenced work on a scaled-up, road-going version of their in-line 4, very few people could have envisaged the impact this machine was to create, and the trend it was to start. This 750 cc machine was designed to be, and indeed became, the number one 'Superbike'. The machine, designated the CB750, followed the lines of the racing machines closely in engine layout, with four separate carburettors and four exhaust pipes and silencers. It was an instant success; in just over three years, in the USA alone, over 61,000 Honda 750 cc fours were sold.

With the overwhelming success of the CB750/4 and the experience gained from its manufacture and production, and the popularity of motor cycles increasing, once again, Honda designed and produced the first of several smaller capacity road-going fours; the CB500/4. This new model, well engineered and pleasantly compact, struck a happy medium between a sports and touring machine. In 1974, probably due to the dictates of a market trend rather than any particular engineering reason, the CB500 had its engine over-bored and was re-designated the CB550/4. It was later to be offered in two different guises, the CB550 K and the CB550 F, ostensibly the same machine in either touring or sporting trim.

The next new model on the same theme, the CB350/4, introduced onto the US market shortly after the 500/4, was destined to be Honda's only failure to date in the four-cylinder field. The machine was under-powered and over-styled and never caught on. It was not marketed in the UK.

Honda seemed determined to "get it right" with their next attempt; the CB400/4. This machine was compact, relatively light, sportingly styled with no unnecessary embellishments, had adequate power and handled extremely well. It possessed all the attributes deemed necessary by many riders, particularly those of a sporting disposition. It was especially popular in the UK where many enthusiasts still mourn its passing.

With the re-emergence of the 650 cc class as a popular choice with riders requiring a competent, all-round machine, Honda have increased the engine capacity of the CB550/4 to create the 626 cc, CB650/4. This model, first introduced into the US and UK in 1979, is mechanically similar to its forerunners although incorporating numerous modifications and detail refinements, and a new frame and cycle parts. The original CB650 Z model is still available in the UK at the time of writing. In the US, the original model has been superseded by two variants for 1980. The CB650 A and CB650 CA models feature identical engines and share many running gear components. The differences between the models occur in the styling. The CB650 A has more traditional styling, retaining wire-spoked wheels and a compact four-into-two exhaust system. The CB650 CA is Honda's custom variant of the CB650. It features reversed Comstar wheels, the rear being of 16 inch diameter, slightly extended front forks and a four silencer exhaust system.

Reference is made in the text, in the dismantling and reassembly procedures, to the various differences in each model's equipment. For information on the later model changes, refer to Chapter 7.

Model dimensions and weights

	CB650 Z (UK)	CB650 Z (US)	CB650 A	CB650 CA
Overall length	2200 mm (87.4 in)	2170 mm (85.4 in)	2180 mm (85.8 in)	2215 mm (87.2 in)
Overall width	775 mm (30.5 in)	850 mm (33.5 in)	865 mm (34.1 in)	870 mm (34.3 in)
Overall height	1110 mm (43.7 in)	1175 mm (46.3 in)	1155 mm (45.5 in)	1155 mm (45.5 in)
Wheelbase	1430 mm (56.3 in)	1430 mm (56.3 in)	1450 mm (57.1 in)	1480 mm (58.3 in)
Seat height	795 mm (31.3 in)	800 mm (31.5 in)	765 mm (30.1 in)	765 mm (30.1 in)
Ground clearance	155 mm (6.1 in)	160 mm (6.3 in)	155 mm) (6.1 in)	145 mm (5.7 in)
Dry weight	198 kg (436 lb)	196 kg (431 lb)	199 kg (437 lb)	201 kg (442 lb)

Ordering Spare Parts

When ordering spare parts for any Honda model it is advisable to deal direct with an official Honda agent, who should be able to supply most items ex-stock. Parts cannot be obtained from Honda (UK) Limited direct: all orders must be routed via an approved agent, even if the parts required are not held in stock.

Always quote the engine and frame numbers in full, particularly if parts are required for any of the earlier models.

The frame number is located on the left hand side of the steering head and the engine number is stamped on the upper crankcase, immediately to the rear of the two left-hand cylinders. Use only parts of genuine Honda manufacture. Pattern parts are available, some of which originate from Japan, but in many instances they may have an adverse effect on performance and/or reliability. Furthermore the fitting on non-standard parts may invalidate the warranty. Honda do not operate a 'service exchange' scheme.

Some of the more expendable parts such as sparking plugs, bulbs, tyres, oils and greases etc., can be obtained from accessory shops and motor factors, who have convenient opening hours, and can often be found not far from home. It is also possible to obtain parts on a Mail Order basis from a number of specialists who advertise regularly in the motor cycle magazines.

Frame number location

Engine number location

Safety first!

Professional motor mechanics are trained in safe working procedures. However enthusiastic you may be about getting on with the job in hand, do take the time to ensure that your safety is not put at risk. A moment's lack of attention can result in an accident, as can failure to observe certain elementary precautions.

There will always be new ways of having accidents, and the following points do not pretend to be a comprehensive list of all dangers; they are intended rather to make you aware of the risks and to encourage a safety-conscious approach to all work you carry out on your vehicle.

Essential DOs and DON'Ts

DON'T start the engine without first ascertaining that the transmission is in neutral.

DON'T suddenly remove the filler cap from a hot cooling system – cover it with a cloth and release the pressure gradually first, or you may get scalded by escaping coolant.

DON'T attempt to drain oil until you are sure it has cooled sufficiently to avoid scalding you.

DON'T grasp any part of the engine, exhaust or silencer without first ascertaining that it is sufficiently cool to avoid burning you.

DON'T allow brake fluid or antifreeze to contact the machine's paintwork or plastic components.

DON'T syphon toxic liquids such as fuel, brake fluid or antifreeze by mouth, or allow them to remain on your skin.

DON'T inhale dust – it may be injurious to health (see *Asbestos* heading).

DON'T allow any spilt oil or grease to remain on the floor – wipe it up straight away, before someone slips on it.

DON'T use ill-fitting spanners or other tools which may slip and cause injury.

DON'T attempt to lift a heavy component which may be beyond your capability – get assistance.

DON'T rush to finish a job, or take unverified short cuts.

DON'T allow children or animals in or around an unattended vehicle.

DON'T inflate a tyre to a pressure above the recommended maximum. Apart from overstressing the carcase and wheel rim, in extreme cases the tyre may blow off forcibly.

DO ensure that the machine is supported securely at all times. This is especially important when the machine is blocked up to aid wheel or fork removal.

DO take care when attempting to slacken a stubborn nut or bolt. It is generally better to pull on a spanner, rather than push, so that if slippage occurs you fall away from the machine rather than on to it.

DO wear eye protection when using power tools such as drill, sander, bench grinder etc.

DO use a barrier cream on your hands prior to undertaking dirty jobs – it will protect your skin from infection as well as making the dirt easier to remove afterwards; but make sure your hands aren't left slippery. Note that long-term contact with used engine oil can be a health hazard.

DO keep loose clothing (cuffs, tie etc) and long hair well out of the way of moving mechanical parts.

DO remove rings, wristwatch etc, before working on the vehicle – especially the electrical system.

DO keep your work area tidy – it is only too easy to fall over articles left lying around.

DO exercise caution when compressing springs for removal or installation. Ensure that the tension is applied and released in a controlled manner, using suitable tools which preclude the possibility of the spring escaping violently.

DO ensure that any lifting tackle used has a safe working load rating adequate for the job.

DO get someone to check periodically that all is well, when working alone on the vehicle.

DO carry out work in a logical sequence and check that everything is correctly assembled and tightened afterwards.

DO remember that your vehicle's safety affects that of yourself and others. If in doubt on any point, get specialist advice.

IF, in spite of following these precautions, you are unfortunate enough to injure yourself, seek medical attention as soon as possible.

Asbestos

Certain friction, insulating, sealing, and other products – such as brake linings, clutch linings, gaskets, etc – contain asbestos. *Extreme care must be taken to avoid inhalation of dust from such products since it is hazardous to health.* If in doubt, assume that they *do* contain asbestos.

Fire

Remember at all times that petrol (gasoline) is highly flammable. Never smoke, or have any kind of naked flame around, when working on the vehicle. But the risk does not end there – a spark caused by an electrical short-circuit, by two metal surfaces contacting each other, by careless use of tools, or even by static electricity built up in your body under certain conditions, can ignite petrol vapour, which in a confined space is highly explosive.

Always disconnect the battery earth (ground) terminal before working on any part of the fuel or electrical system, and never risk spilling fuel on to a hot engine or exhaust.

It is recommended that a fire extinguisher of a type suitable for fuel and electrical fires is kept handy in the garage or workplace at all times. Never try to extinguish a fuel or electrical fire with water.

Note: *Any reference to a 'torch' appearing in this manual should always be taken to mean a hand-held battery-operated electric lamp or flashlight. It does **not** mean a welding/gas torch or blowlamp.*

Fumes

Certain fumes are highly toxic and can quickly cause unconsciousness and even death if inhaled to any extent. Petrol (gasoline) vapour comes into this category, as do the vapours from certain solvents such as trichloroethylene. Any draining or pouring of such volatile fluids should be done in a well ventilated area.

When using cleaning fluids and solvents, read the instructions carefully. Never use materials from unmarked containers – they may give off poisonous vapours.

Never run the engine of a motor vehicle in an enclosed space such as a garage. Exhaust fumes contain carbon monoxide which is extremely poisonous; if you need to run the engine, always do so in the open air or at least have the rear of the vehicle outside the workplace.

The battery

Never cause a spark, or allow a naked light, near the vehicle's battery. It will normally be giving off a certain amount of hydrogen gas, which is highly explosive.

Always disconnect the battery earth (ground) terminal before working on the fuel or electrical systems.

If possible, loosen the filler plugs or cover when charging the battery from an external source. Do not charge at an excessive rate or the battery may burst.

Take care when topping up and when carrying the battery. The acid electrolyte, even when diluted, is very corrosive and should not be allowed to contact the eyes or skin.

If you ever need to prepare electrolyte yourself, always add the acid slowly to the water, and never the other way round. Protect against splashes by wearing rubber gloves and goggles.

Mains electricity and electrical equipment

When using an electric power tool, inspection light etc, always ensure that the appliance is correctly connected to its plug and that, where necessary, it is properly earthed (grounded). Do not use such appliances in damp conditions and, again, beware of creating a spark or applying excessive heat in the vicinity of fuel or fuel vapour. Also ensure that the appliances meet the relevant national safety standards.

Ignition HT voltage

A severe electric shock can result from touching certain parts of the ignition system, such as the HT leads, when the engine is running or being cranked, particularly if components are damp or the insulation is defective. Where an electronic ignition system is fitted, the HT voltage is much higher and could prove fatal.

Routine Maintenance

For information relating to 1981-on models refer to Chapter 7

Periodic routine maintenance is essential to keep the motorcycle in a safe condition and at peak performance. Routine maintenance also saves money because it provides the opportunity to detect and remedy a fault before it develops further and causes more damage. Maintenance should be undertaken on either a calendar or mileage basis depending on whichever comes sooner. The period between maintenance tasks serves only as a guide since there are many variables, eg: age of machine, riding technique and adverse conditions.

The maintenance instructions are generally those recommended by the manufacturer but are supplemented by additional tasks which, through practical experience, the author recommends should be carried out at the intervals suggested. The additional tasks are primarily of a preventative nature, which will assist in eliminating unexpected failure of a component or system, due to wear and tear, and increase safety margins when riding.

All the maintenance tasks are described in detail together with the procedures required for accomplishing them. If necessary, more general information on each topic can be found in the relevant Chapter within the main text.

Although no special tools are required for routine maintenance, a good selection of general workshop tools is essential. Included in the tools must be a range of metric ring or combination spanners, a selection of crosshead screwdrivers, and two pairs of circlip pliers, one external opening and the other internal opening. Additionally, owing to the extreme tightness of most casing screws on Japanese machines, an impact screwdriver, together with a choice of large or small cross-head screw bits, is absolutely indispensable. This is particularly so if the engine has not been dismantled since leaving the factory.

Weekly or every 200 miles (320 km)

1 Tyres

Check the tyre pressures. Always check the pressure when the tyres are cold as the heat generated when the machine has been ridden can increase the pressure by as much as 8 psi, giving a totally inaccurate reading. Variations in pressure of as little as 2 psi may alter certain handling characteristics. It is therefore recommended that whatever type of pressure gauge is used, it should be checked occasionally to ensure accurate readings. Do not put absolute faith in 'free air' gauges at garages or petrol stations. They have been known to be in error. Inspect the tyre treads for cracking or evidence that the outer rubber is leaving the inner cover. Also check the sidewalls for splitting or perishing. Remove any stones or other road debris which may be embedded between the threads. Remove any such objects with a suitable tool such as small screwdriver.

When checking the tyres for damage, they should be examined for tread depths in view of both the legal and safety aspects. It is vital to keep the tread depth above the UK legal limit of 1 mm of depth over three-quarters of the tread breadth and around the entire circumference. Many riders, however, consider nearer 2 mm to be the limit for secure road-holding, traction and braking, especially when riding in adverse weather and road conditions. Honda themselves recommend replacement of the front tyre when there is 1.5 mm (0.06 in) of tread remaining, and replacement of the rear tyre when wear has reduced the tread depth to 2.0 mm (0.08 in).

2 Engine oil

Check the engine oil level by means of the dipstick incorporated in the filler plug which screws into the left-hand side of the crankcase. When taking the reading do not screw the plug into the casing: allow it to rest on the rim of the filler orifice. Replenish the engine oil with oil of the specified grade to the maximum level on the dipstick.

3 Final drive chain: interim inspection and lubrication

Examine the final drive chain for excessive play or lack of lubrication. If either of these defects are apparent, they should be remedied by referring to the fully detailed section under the Monthly (800 miles) heading. It is difficult to give any specific intervals for adjustment and lubrication because there are so many variables which bear on the rate of chain wear, not least of which is the 'enthusiasm' which the rider displays in his riding style.

4 Electrical system

Check that the various bulbs are functioning properly, paying particular attention to the rear lamp. It is possible that one of the rear lamp or brake filaments has failed but gone unnoticed. Check that the indicators and horn operate normally. Clean all lenses. If any of the fuses has blown recently, check that the source of the problem has been resolved and that the spare fuse has been renewed.

5 Safety inspection

Give the whole machine a close visual inspection, checking for loose nuts and fittings, frayed control cables and damaged brake hoses etc.

6 Control cable intermediate lubrication

Apply a small amount of light oil to the exposed ends of the control inner cables. This will prevent drying-up of the cables between the more thorough lubrication which should be given every four months or 4000 miles.

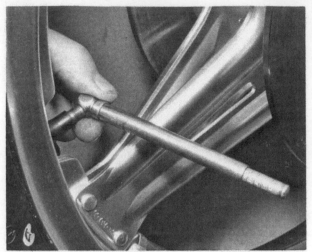

Use pocket pressure gauge for regular tyre pressure checks

Check engine oil with dipstick

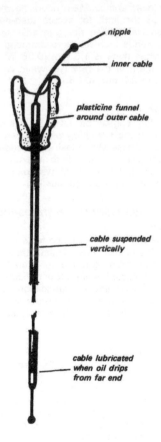

nipple

inner cable

plasticine funnel
around outer cable

cable suspended
vertically

cable lubricated
when oil drips
from far end

Oiling a control cable

Monthly or every 800 miles (1320 km)

Complete all the checks listed under the previous maintenance interval heading and then carry out the following:

1   Final drive chain: cleaning and lubrication

The exact interval at which the final drive chain will require lubrication is entirely dependent on the use to which the machine is put. In some cases the chain will require daily lubrication, in other cases it need only be lubricated at weekly intervals. The best rule to follow is that if the chain rollers look dry, then the chain needs lubrication immediately. Do not allow the chain to run dry until the links start to kink, or until traces of reddish-brown deposit can be seen on the sideplates.

Note that two types of chain are available, standard chains and O-ring chains. O-ring chains are wider and have small O-rings set between the inner and outer sideplates of each link to seal the grease into the bearings for the life of the chain. These are easily identified by the presence of the O-rings. Although fitted as standard on later models, O-ring chains are increasingly popular due to their increased life and may be encountered on earlier models with replacement chains.

On standard chains, frequent lubrication is best carried out with one of the aerosol-applied chain greases. It is, however, recommended that at longer intervals, between 500-1000 miles, the chain be removed, cleaned and lubricated in a chain bath, such as Linklyfe or Chainguard.

On O-ring chains, on no account subject the chain to a chain bath, such treatment would damage the O-rings and destroy the chain's internal self lubricating qualities. Apart from paraffin, do not use solvents to clean the chain, these too will damage the O-rings. Honda recommend the use of SAE 80 or 90 gear oil to lubricate the chain, although one of the aerosol type chain lubricants may be used if it is marked as being suitable for use on O-ring chains.

In particularly adverse weather conditions, or when touring, lubrication should be undertaken more frequently.

A final word of caution; the importance of chain lubrication cannot be overstressed in view of the cost of replacement, and the fact that a considerable amount of dismantling work, including swinging arm removal, will need to be undertaken should replacement be necessary.

2   Final drive chain: adjustment

Adjust the chain after lubrication, so that there is approximately 15-25 mm ($\frac{5}{8}$-1 inch) slack in the middle of the lower run. Always check with the chain at the tightest point as a chain rarely wears evenly during service.

Adjustment is accomplished after placing the machine on the centre stand and slackening the wheel nut, so that the wheel can be drawn backwards by means of the drawbolt adjusters in the fork ends.

The torque arm nuts and the rear brake adjuster must also be slackened during this operation. Adjust the drawbolts an equal amount to preserve wheel alignment. The fork ends are clearly marked with a series of parallel lines above the adjusters, to provide a simple visual check.

With the adjustment correct, tighten the wheel nut and fit a new split pin. A good investment here is a large R-clip; this performs the function of the split pin and may be re-used time and time again. It is also much easier to withdraw and refit, especially when the spindle nut is obscured by two exhaust silencers.

3   Brake fluid

Check the hydraulic fluid level in the front brake master cylinder reservoir. Before removing the reservoir cap and diaphragm place the handlebars in such a position that the reservoir is approximately vertical. This will prevent spillage. The fluid should lie between the upper and lower lines on the reservoir body. Replenish, if necessary, with hydraulic brake fluid of the correct specification, which is DOT 3 (USA) or SAE-J1703. If the level of fluid in the reservoir is excessively low, check the pads for wear. If the pads are not worn, suspect a fluid leakage in the system. This must be rectified immediately.

4   Brake wear

Check that when applied, the rear brake wear indicator is within the usable range scale marked on the brake plate. The front disc pads should also be examined for wear, and to this end are marked with a red line denoting the maximum wear limit. If necessary, change the pads and/or brake shoes, referring to Chapter 5 for details. Look also for signs of staining on the friction material. This may be caused by leakage from the fork leg or from the caliper seals; in either case attention must be given to locating and rectifying the source of the leak.

5   Battery

Maintenance of the battery is normally limited to keeping the electrolyte level correct.

Check the electrolyte level in the battery and replenish, if necessary, with distilled water. Do not use tap water as this will reduce the life of the battery. If the battery is removed for filling, note the tracking of the battery breather pipe which should be replaced in the same position, ensuring that the pipe is not kinked or blocked. If the breather pipe is restricted and the battery overheats for any reason, the pressure produced may, in extreme cases, cause the battery case to fail and a liberal amount of sulphuric acid to be deposited on the electrical harness and frame parts.

Brake fluid **must not** fall below lower level line

Maintain electrolyte level between upper and lower level lines

6   Further maintenance checks

The following areas should be given a cursory check, taking remedial action where required. Check the electrical system, plus the headlamp beam alignment. Check the various nuts, bolts and screws for security, tightening where necessary. Check the front and rear suspension for smooth operation. Check the steering head bearings for free play. Examine all control cables and hydraulic lines, renewing any which appear worn or frayed.

Four Monthly or every 4000 miles (6400 km)

Complete the checks in the preceding maintenance schedules and then carry out the following:

1   Engine oil and filter renewal

Run the engine until normal operating temperature is reached to ensure that the old oil drains quickly and completely. Place the machine on its centre stand and position a drain tray or bowl of about 1 gallon capacity beneath the sump drain plug. Release the drain bolt and filler plug, and allow the engine oil to drain.

Slacken the oil filter bolt and remove the filter housing and element. Note that some residual oil will be released, and some provision must be made to catch this. When all the oil has drained, clean the area around the drain plug and filter housing, and the inside of the filter bowl. Refit and tighten the drain plug after checking that the sealing washer is in place and not damaged. No attempt should be made to clean the old filter; it must be discarded and a new component fitted. When fitting a new filter, install first the filter spring into the chamber followed by the washer, the filter element and then the spacing collar. Check the condition of the O-rings on the filter bolt and mating face of the housing, before refitting the housing and tightening the bolt. Do not overtighten the filter housing bolt and use a close fitting spanner when releasing and tightening the bolt.

Fill the crankcase with 3.0 litres (5.2 Imp. pints/3.2 US quarts) of the recommended engine oil, then run the engine for two or three minutes, checking for signs of leakage around the drain plug and filter. Stop the engine and check the oil level, adding oil where necessary to bring the level to maximum on the dipstick.

2   Valve clearances

It is important that valve clearances be maintained otherwise damage, or at best poor performance and noisy operation, will occur. To gain access to the tappets, raise the dualseat and remove the petrol tank as described in Chapter 2, Section 2.

Remove the right- and left-hand cylinder head inspection covers, each of which is retained by two bolts, and the central breather cover. Detach the large breather hose from the cover prior to removing it. The central breather cover is retained by two bolts. Remove the circular inspection cover from the right-hand end of the crankcase, to expose the pulser generator assembly.

Note that the engine must be **cold** during the valve clearance check. Using a spanner on the hexagonal spacer fitted outboard of the pulser generator assembly, turn the engine clockwise (forwards) and watch the inlet valve tappet of No 1 cylinder (extreme left-hand cylinder). Because the engine turning hexagon is on the right-hand side of the engine and inspection is required on the left-hand side at this stage, the use of an assistant to turn the engine can be most beneficial. Continue turning the engine until the inlet valve has opened and then closed. Then align the timing mark '1.4T', viewed through the aperture in the pulser generator baseplate, so that it is exactly in line with the fixed timing mark pointer. No 1 piston is now exactly at top dead centre (TDC) and both valves are fully closed (both tappets should have free play). The valve clearances of No 1 cylinder can now be checked by means of a feeler gauge as can the inlet valve of No 3 cylinder and the exhaust valve of No 2 cylinder – which are also closed in this engine position.

The standard valve clearance is 0.05 mm (0.002 in) for the inlet valves and 0.08 mm (0.003 in) for the exhaust valves.

Adjustment is made by slackening the locknut at the end of the rocker arm and by turning the adjusting screw until the feeler gauge is a good sliding fit. Hold the adjusting screw whilst the locknut is retightened, then recheck the valve clearance.

Rotate the engine one complete turn (360°) until the piston in No 4 cylinder is exactly at top dead centre, by following an identical procedure. Both valves on No 4 cylinder can be checked, also the remaining valves on the other cylinders – inlet valve on No 2 and the exhaust valve on No 3. Recheck the settings after adjustment and tightening the locknuts.

It is most important that the valve clearances are checked when the valves are fully closed. If adjustments are made when a valve is still slightly on lift, an excessive clearance will result, making the valve gear noisy and causing loss of performance. NEVER set the clearances less than the recommended values. If the valves do not seat correctly when the engine is hot, both valves and seatings will burn away, necessitating an expensive top overhaul of the engine. It is never practicable to quieten a noisy engine in this manner. If the valve clearances are checked when the engine is in the positions indicated, this will ensure the valves concerned are fully closed.

Refit the individual inspection covers for Nos 1 and 4 cylinder and the central breather cover. Ensure all the rubber sealing gaskets are in sound condition. Reconnect the breather hose to the cover. Refit the fuel tank. The pulser generator inspection cover should not be refitted until completing the following adjustment.

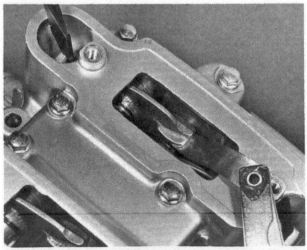

Drain plug for engine/transmission oil

Replenish engine oil through aperture in top of casing

Adjust valve clearances with inspection covers and breather cover removed. Engine must be **cold**

Align timing mark 1.4T with fixed pointer as described

3 Cam chain tensioner adjustment

The camshaft drive chain is tensioned by a spring blade assembly mounted in the rear of the cam chain tunnel. The tensioner is semi-automatic, adjusting to the correct tension when the locknut is released. Note that the adjustment procedure should be carried out with the engine stopped.

Slacken the small domed locknut at the lower rear of the cam chain tunnel. Using a spanner on the hexagonal spacer fitted outboard of the pulser generator assembly, slowly rotate the crankshaft clockwise (forwards) and, at the same time, retighten the locknut. The adjustment is now correct. Refit the inspection cover.

Use domed nut for cam chain tensioner adjustment

4 Sparking plugs: cleaning and checking the gaps

Remove, clean and adjust the sparking plugs. Carbon and other deposits can be removed, using a wire brush, and emery paper or a file used to clean the electrodes prior to adjusting the gaps. Probably the best method of sparking plug cleaning is by having them shot blasted in a special machine. This type of machine is used by most garages. If the outer electrode of a plug is excessively worn (indicated by a step in the underside) the plug should be renewed. Adjust the points gap on each plug by bending the outer electrode only, so that the gap is within the range 0.6 – 0.7 mm (0.024 – 0.028 in). Before replacing the plugs, smear the threads with graphited grease; this will aid subsequent removal. Whenever the plugs are removed, make a visual inspection of the condition of the plugs, especially the colour of the carbon deposits.

If the standard plug appears deficient; either appearing to be too hot (too soft a grade) or too cold (too hard a grade), then alternative plug types should be installed.

Note that the advice of an authorized Honda Service Agent or similar expert should be sought before the plug heat range is altered from standard. The use of too cold, or hard, a grade of plug will result in fouling and the use of too hot, or soft, a grade of plug will result in engine damage due to the excess heat being generated.

5 Air filter: cleaning the element

Access to the filter element is gained after removing the left-hand side panel on CB650 Z models and the right-hand side panel on the CB650 A and CA models. Remove the air filter case cover by releasing the two or three screws, (depending on the model) which retain it. The element can now simply be pulled clear of the case.

Tap the element gently to remove any loose dust and then use an air hose to remove the remainder of the dust. Apply the air current from the inside of the element only. If an air hose is not available, a tyre pump can be utilised instead. If the corrugated paper element is damp, oily or beginning to disintegrate, it must be renewed. Do not run the engine with the element removed as the weak mixture caused may result in engine overheating and damage to the cylinders and pistons.

Ensure the rubber sealing rings on the case covers are refitted renewing them if perished or split; a bad joint here can also result in a weak mixture.

On US models it will also be necessary to drain the crankcase breather tube. Simply remove the plug from the end of the drain tube situated just to the rear of the centre stand pivot and allow the contents of the tube to drain into a suitable container. Once the contents of the tube have drained refit the plug to the end of the tube and secure it in position with its clamp.

6 Carburettors: adjustment and synchronisation

The following points should be checked, and if necessary, adjusted. Note that adjustments should not be made by way of experimentation – if all is well, leave the carburettors alone. In practice it will be found that carburation adjustments are maintained with reasonable accuracy over quite long periods.

Run the engine to raise the temperature to normal, preferably by riding the machine for 10 – 15 minutes. Place the machine on its centre stand and allow the engine to idle. Check the idle speed, which should be 1050 ± 100 rpm. If necessary, move the throttle stop screw to bring the idle speed within limits.

Carburettor synchronisation requires the use of a vacuum gauge set. If this is not available, do not attempt adjustment, but take the machine to a Honda dealer to have this operation carried out.

If the vacuum gauge set is available, proceed as follows. Remove the dualseat and petrol tank so that access can be gained to the carburettors. Using a suitable length of feed pipe, reconnect the petrol tank with the carburettors, so that the petrol flow can be maintained. The petrol tank must be placed above the level of the carburettors. Remove the blanking plugs from the take-off points on the inlet tracks to the rear of the cylinder head, and connect the vacuum gauges to the engine.

Start the engine and allow it to run until normal working temperature has been reached. This should take 10 – 15 minutes. Set the throttle so that an engine speed of 1050 ± 100 rpm is maintained. If the readings on the vacuum gauges vary by more than 60 mm Hg (2.4 in Hg) it will be necessary to bring the carburettors within limits using the bell crank adjuster screws fitted to the top of carburettors No 1, 3 and 4. Remove the two screws and remove the tops from each of these carburettors. Note that if the readings on the gauge fluctuate wildly, it is likely that the gauges require heavier damping. Refer to the gauge manufacturer's instructions on setting up procedures.

The No 2 carburettor (second from the left) is regarded as the base instrument; that is, it is non-adjustable and the remaining three carburettors must be adjusted to it. Honda produce a special combined screwdriver and socket spanner for dealing with the bell crank adjuster screws (Part Number 07908-4220110). Its use makes the procedure easier, but it is not essential. Slacken the locknut of the adjuster concerned, then turn the latter, noting the effect on the gauge reading. When the reading is as close as possible to that of the No 2 carburettor, hold the adjuster screw and retighten the locknut. Repeat the procedure on the remaining carburettors. Ensure all the locknuts are secure. Disconnect the vacuum gauges and refit the inlet track take-off point blanking plugs. Refit and secure the carburettor tops.

To check the operation of the throttle twistgrip and cables, turn the fuel supply off and allow the engine to idle until it stalls due to the lack of fuel. This will prevent subsequent flooding problems with the accelerator pump fitted to the No 2 cylinder carburettor.

The throttle twistgrip should have about 2 – 6 mm free play measured at the inner, flanged, edge. Coarse adjustment can be made by moving the lower adjuster by the required amount. Further fine adjustment is made by means of the upper adjuster.

Use throttle stop screw to adjust idle speed

7   Wheel condition – wire spoked types

Check the spoke tension by gently tapping each one with a metal object. A loose spoke is identifiable by the low pitch noise generated. If any spoke needs considerable tightening, it will be necessary to remove the tyre and inner tube in order to file down the protruding spoke end. This will prevent the spoke from chafing through the rim band and piercing the inner tube. Rotate the wheel and test for rim runout. Excessive runout will cause handling problems and should be corrected by tightening or loosening the relevant spokes. Care must be taken, since altering the tension in the wrong spokes may create more problems.

8   Side stand: pad renewal

The side stand is fitted with a rubber pad which will gradually wear down with use. Check the pad condition and renew it when it nears the raised wear line. The old pad can be released by removing the single retaining bolt and the new item fitted by reversing the dismantling sequence. Ensure only pads bearing the inscription 'over 260 lbs only' are fitted.

Ensure that the side stand return spring retains the stand securely in the retracted position. If this is not the case the spring must be renewed. Apply a small amount of grease to the side stand pivot to prevent corrosion and ensure the stand operates smoothly.

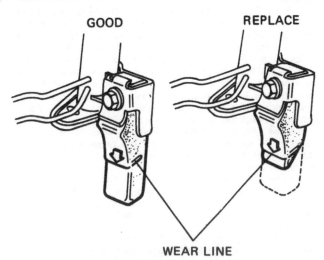

GOOD   REPLACE

WEAR LINE

Sidestand rubber wear limit

9   Swinging arm bushes – lubrication

In order to maintain correct swinging arm operation, and prevent premature swinging arm bush failure, the bush bearings must be regularly lubricated. A grease nipple located on the swinging arm cross member is provided for this purpose. Pump grease into the nipple using a grease gun with a suitably shaped extension to gain access to the grease nipple. Wipe away any excess grease from around the nipple when lubrication is finished.

10   Control cable lubrication

Use motor oil or an all-purpose oil to lubricate the control cables. A good method for lubricating the cables is shown in the accompanying illustration, using a plasticine funnel. This method has the disadvantage that the cables usually need removing from the machine. An hydraulic cable oiler which pressurises the lubricant, forcing it along the cable, overcomes this problem. Nylon lined cables should not be lubricated; in some cases the lubricant will cause the cable lining to swell leading to seizure.

Eight monthly or every 8000 miles (12 800 km)

Carry out the operations listed in the previous Sections, then carry out the following.

1   Renew the sparking plugs
2   Renew the air filter element

In addition, the machine should be given a close visual examination, checking the numerous details not covered by the normal maintenance schedule. Check for signs of corrosion or rusting around the frame and cycle parts, taking the appropriate remedial action where required.

Additional routine examination

1   Brake pads: examination and replacement

The rate of brake pad wear is dependent on the conditions under which the machine operates, weight carried and the style of riding, consequently it is difficult to advise on specific inspection intervals. Whatever inspection interval is chosen bear in mind that the rate of wear will not be constant.

To check wear on the front brake pads examine the pads through the small window in the main caliper units. If the red mark of any pad has been reached, both pads in that set must be renewed. The rate of wear of the two sets are similar so it is probable that they will require renewal at the same time in any case.

2   Brake pad renewal

Remove each brake set individually, using an identical procedure as follows:

Unscrew the two bolts that pass into the caliper body and secure the body to the support bracket. Lift the caliper body off the support bracket, still interconnected with the hydraulic hose.

Lift the old pads out. Install the new pads and also the shim which fits against the outer face of the outer pad. The shim must be fitted so that the arrow is in the forward-most position, pointing in an upward direction. Refit the caliper halves and replace the socket screws. It may be necessary to push the caliper cylinder piston inwards to give the necessary clearance. If required, the bleed screw on the caliper can be slackened at the same time as the piston is pushed inwards. This will allow a small amount of fluid to seep out and the piston to move. Place a rag around the bleed screw to prevent the fluid leaking onto the caliper unit. Operate the brake lever, after pad replacement, to check free movement of the pads and to allow the pads to self-adjust.

Brake pads can be inspected through small window in caliper

3 Rear brake adjustment

Adjustment of the rear brake should be carried out when required, the intervals being dependent on riding style and usage. The adjustment is carried out by turning the nut on the rear end of the brake operating rod. Precise adjustment is a matter of personal choice but it should be ensured that the brake does not bind when the foot pedal is in the fully returned position. A brake wear indicator is fitted to the brake actuating arm on the back plate. If, when the brake is applied fully, the arrow on the arm is in line with the cast-in index mark on the back plate, the brake linings are worn sufficiently to require renewal.

4 Clutch adjustment

In common with brake pad wear, clutch wear and the resultant necessary adjustment depends on operating conditions and the style of riding. Adjust the clutch, when necessary, as follows.

Check the clutch free play at the ball-end of the handlebar lever. The lever should move 10 – 20 mm ($\frac{3}{8}$ – $\frac{3}{4}$ in) before the clutch begins to lift. If the free play is incorrect, the cable may be adjusted by means of the adjuster screws at both ends of the cable. The lower adjuster is used for coarse adjustments and the upper adjuster for finer running adjustments. If the upper

adjuster threads project more than 8 mm (0.3 in) from the lever stock, there is some danger of the adjuster breaking out of the stock boss. To prevent this adjustment must be made at the clutch housing. Slacken the lower cable adjuster locknut and, using the adjuster nut, obtain maximum free play. Remove the clutch housing central inspection cap, and slacken the clutch release arm locknut which is now visible. Turn the adjuster screw clockwise until a slight resistance is felt. From this position, turn the adjusting screw anticlockwise $\frac{3}{4}$ of a turn, and retighten the locknut. Refit the inspection cap. Using the cable lower adjuster nut, adjust the cable until there is 10 – 20 mm ($\frac{3}{8}$ – $\frac{3}{4}$ in) of free play at the ball end of the clutch lever. Retighten the lower adjuster locknut. Any further fine running adjustment can now be made at the upper cable adjuster at the handlebar lever.

5 Front fork oil renewal

When the action of the front forks becomes soft or spongy (ie diving too much under heavy front wheel braking) or the damping feels weak, then the fork oil is due for replacement. Note that these symptoms can also be caused by worn or damaged internal components, but by changing the fork oil prior to investigative dismantling, a lot of unnecessary work may be avoided.

Prise out the caps from the fork top bolts. Unscrew and remove the large socket head top bolts. This is not easy with the fork legs in their normal position, because the handlebars obstruct the bolts. The easiest method is to slacken the pinch bolts in the upper and lower yoke, slide the stanchion down about 3 inches, then tighten the lower pinch bolts. Slacken and remove the top bolt using the appropriate hexagon socket key. If this is unavailable, a suitable alternative can be found in a hexagon head bolt of the required size, inserted into a socket spanner. Release the top bolts carefully. Do not forget to remove the spring top seats, fitted below the top bolts, prior to refilling.

Place a suitable container below the drain bolt to the rear of each lower leg and remove each bolt. Allow the fork oil to drain, compressing the suspension slightly to drain the oil fully. Re-install the drain bolts, checking that the copper sealing washers are refitted, and refill the forks, using ATF (Automatic Transmission Fluid). Do not overfill. Refit and tighten the fork top bolt in the reverse manner of that described for removal.

6 Brake fluid renewal

Renew the brake fluid every two years to preserve maximum braking efficiency by ensuring that the fluid has not been contaminated and deteriorated to an unsafe degree.

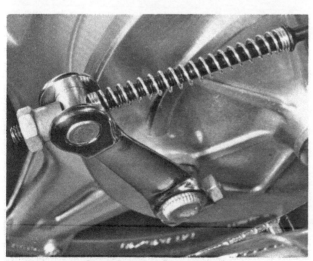

Use brake operating rod nut to adjust rear brake

Clutch cable lower adjuster is used for coarse adjustment

Further clutch adjustment requires removal of the inspection cap and ...

... adjustment of the screw with the locknut slackened

Before starting work, obtain a new, full can of the recommended brake fluid and read carefully the Section on brake bleeding in Chapter 5. Prepare the plastic pipe and glass jar in the same way as for bleeding, unscrew the bleed nipple ¼ – ½ a turn and apply the front brake lever gently and repeatedly. This will pump out the old fluid. *Keep the master cylinder topped up at all times, otherwise air may enter the system and greatly lengthen the operation.* The old brake fluid is invariably much darker in colour than the new, making it easy to see when it has been pumped out and the new fluid has completely replaced it.

When the new fluid appears in the clear plastic tube completely uncontaminated by traces of old fluid, close the bleed nipple and replace the rubber cap on the nipple. Top up the master cylinder reservoir above the lower level mark, unless this operation has been combined with brake pad renewal, in which case slightly more fluid should be added. Clean and dry the rubber diaphragm, fold it into its compressed state and refit the diaphragm and reservoir cover, tightening its retaining screws securely.

Wash off any surplus fluid and check that the brake is operating correctly before taking the machine out on the road.

Remove the drain bolts to drain fork oil

Standard torque settings

Specific torque settings will be found at the end of the specifications section of each chapter. Where no figure is given, bolts should be secured according to the table below.

Fastener type (thread diameter)	kgf m	lbf ft
5 mm bolt or nut	0.45 – 0.6	3.5 – 4.5
6 mm bolt or nut	0.8 – 1.8	6 – 9
8 mm bolt or nut	1.8 – 2.5	13 – 18
10 mm bolt or nut	3.0 – 4.0	22 – 29
12 mm bolt or nut	5.0 – 6.0	36 – 43
5 mm screw	0.35 – 0.5	2.5 – 3.6
6 mm screw	0.7 – 1.1	5 – 8
6 mm flange bolt	1.0 – 1.4	7 – 10
8 mm flange bolt	2.4 – 3.0	17 – 22
10 mm flange bolt	3.0 – 4.0	22 – 29

Quick glance
maintenance adjustments and capacities

Engine/transmission oil capacity
Dry .... 3.5 litre (6.2 Imp pint/3.7 US quart)
At oil/oil filter change .... 3.0 litre (5.2 Imp pint/3.2 US quart)

Sparking plug gap .... 0.6 – 0.7 mm (0.024 – 0.028 in)

Front fork oil capacity (per leg)
Dry:
    CB650 Z .... 170 cc
    CB650 A .... 175 cc
    CB650 CA .... 209 cc
At oil change:
    CB650 Z .... 150 cc
    CB650 A .... 155 cc
    CB650 CA .... 190 cc

Valve clearances (Cold)
Inlet .... 0.05 mm (0.002 in)
Exhaust .... 0.08 mm (0.003 in)

Tyre pressures
Front .... 28 psi (2.0 kg/cm^2)
Rear (solo):
    CB650 Z (UK) .... 28 psi (2.0 kg/cm^2)
    CB650 Z (US), A and CA .... 32 psi (2.25 kg/cm^2)
Rear (with more than 90 kg [200 lb] load):
    CB650 Z (UK/US) .... 40 psi (2.8 kg/cm^2)
    CB650 A .... 36 psi (2.5 kg/cm^2)
    CB650 CA .... 32 psi (2.25 kg/cm^2)

Recommended lubricants

Component	Lubricant
Engine/transmission	
General, all temperature use	SAE 10W/40
Above 15°C (60°F)	SAE 30
-10°C to + 15°C (15° – 60°F)	SAE 20 or 20W
Above -10°C (15°F)	SAE 20W/50
Below 0°C (32°F)	SAE 10W
Front forks	Automatic transmission fluid (ATF)
Final drive chain	
Standard chain	Aerosol chain lubricant
O-ring chain	SAE 80 or 90 gear oil or aerosol lubricant marked as suitable for O-ring chains
General lubrication	Light machine oil
Wheel bearings	High melting point grease
Swinging arm	High melting point grease
Component lubrication during engine reassembly (see text)	Molybdenum disulphide grease

Conversion factors

Length (distance)

	X		=		X		=	
Inches (in)	X	25.4	=	Millimetres (mm)	X	0.0394	=	Inches (in)
Feet (ft)	X	0.305	=	Metres (m)	X	3.281	=	Feet (ft)
Miles	X	1.609	=	Kilometres (km)	X	0.621	=	Miles

Volume (capacity)

	X		=		X		=	
Cubic inches (cu in; in³)	X	16.387	=	Cubic centimetres (cc; cm³)	X	0.061	=	Cubic inches (cu in; in³)
Imperial pints (Imp pt)	X	0.568	=	Litres (l)	X	1.76	=	Imperial pints (Imp pt)
Imperial quarts (Imp qt)	X	1.137	=	Litres (l)	X	0.88	=	Imperial quarts (Imp qt)
Imperial quarts (Imp qt)	X	1.201	=	US quarts (US qt)	X	0.833	=	Imperial quarts (Imp qt)
US quarts (US qt)	X	0.946	=	Litres (l)	X	1.057	=	US quarts (US qt)
Imperial gallons (Imp gal)	X	4.546	=	Litres (l)	X	0.22	=	Imperial gallons (Imp gal)
Imperial gallons (Imp gal)	X	1.201	=	US gallons (US gal)	X	0.833	=	Imperial gallons (Imp gal)
US gallons (US gal)	X	3.785	=	Litres (l)	X	0.264	=	US gallons (US gal)

Mass (weight)

	X		=		X		=	
Ounces (oz)	X	28.35	=	Grams (g)	X	0.035	=	Ounces (oz)
Pounds (lb)	X	0.454	=	Kilograms (kg)	X	2.205	=	Pounds (lb)

Force

	X		=		X		=	
Ounces-force (ozf; oz)	X	0.278	=	Newtons (N)	X	3.6	=	Ounces-force (ozf; oz)
Pounds-force (lbf; lb)	X	4.448	=	Newtons (N)	X	0.225	=	Pounds-force (lbf; lb)
Newtons (N)	X	0.1	=	Kilograms-force (kgf; kg)	X	9.81	=	Newtons (N)

Pressure

	X		=		X		=	
Pounds-force per square inch (psi; lbf/in²; lb/in²)	X	0.070	=	Kilograms-force per square centimetre (kgf/cm²; kg/cm²)	X	14.223	=	Pounds-force per square inch (psi; lbf/in²; lb/in²)
Pounds-force per square inch (psi; lbf/in²; lb/in²)	X	0.068	=	Atmospheres (atm)	X	14.696	=	Pounds-force per square inch (psi; lbf/in²; lb/in²)
Pounds-force per square inch (psi; lbf/in²; lb/in²)	X	0.069	=	Bars	X	14.5	=	Pounds-force per square inch (psi; lbf/in²; lb/in²)
Pounds-force per square inch (psi; lbf/in²; lb/in²)	X	6.895	=	Kilopascals (kPa)	X	0.145	=	Pounds-force per square inch (psi; lbf/in²; lb/in²)
Kilopascals (kPa)	X	0.01	=	Kilograms-force per square centimetre (kgf/cm²; kg/cm²)	X	98.1	=	Kilopascals (kPa)
Millibar (mbar)	X	100	=	Pascals (Pa)	X	0.01	=	Millibar (mbar)
Millibar (mbar)	X	0.0145	=	Pounds-force per square inch (psi; lbf/in²; lb/in²)	X	68.947	=	Millibar (mbar)
Millibar (mbar)	X	0.75	=	Millimetres of mercury (mmHg)	X	1.333	=	Millibar (mbar)
Millibar (mbar)	X	0.401	=	Inches of water (inH₂O)	X	2.491	=	Millibar (mbar)
Millimetres of mercury (mmHg)	X	0.535	=	Inches of water (inH₂O)	X	1.868	=	Millimetres of mercury (mmHg)
Inches of water (inH₂O)	X	0.036	=	Pounds-force per square inch (psi; lbf/in²; lb/in²)	X	27.68	=	Inches of water (inH₂O)

Torque (moment of force)

	X		=		X		=	
Pounds-force inches (lbf in; lb in)	X	1.152	=	Kilograms-force centimetre (kgf cm; kg cm)	X	0.868	=	Pounds-force inches (lbf in; lb in)
Pounds-force inches (lbf in; lb in)	X	0.113	=	Newton metres (Nm)	X	8.85	=	Pounds-force inches (lbf in; lb in)
Pounds-force inches (lbf in; lb in)	X	0.083	=	Pounds-force feet (lbf ft; lb ft)	X	12	=	Pounds-force inches (lbf in; lb in)
Pounds-force feet (lbf ft; lb ft)	X	0.138	=	Kilograms-force metres (kgf m; kg m)	X	7.233	=	Pounds-force feet (lbf ft; lb ft)
Pounds-force feet (lbf ft; lb ft)	X	1.356	=	Newton metres (Nm)	X	0.738	=	Pounds-force feet (lbf ft; lb ft)
Newton metres (Nm)	X	0.102	=	Kilograms-force metres (kgf m; kg m)	X	9.804	=	Newton metres (Nm)

Power

	X		=		X		=	
Horsepower (hp)	X	745.7	=	Watts (W)	X	0.0013	=	Horsepower (hp)

Velocity (speed)

	X		=		X		=	
Miles per hour (miles/hr; mph)	X	1.609	=	Kilometres per hour (km/hr; kph)	X	0.621	=	Miles per hour (miles/hr; mph)

Fuel consumption*

	X		=		X		=	
Miles per gallon, Imperial (mpg)	X	0.354	=	Kilometres per litre (km/l)	X	2.825	=	Miles per gallon, Imperial (mpg)
Miles per gallon, US (mpg)	X	0.425	=	Kilometres per litre (km/l)	X	2.352	=	Miles per gallon, US (mpg)

Temperature

Degrees Fahrenheit = (°C x 1.8) + 32

Degrees Celsius (Degrees Centigrade; °C) = (°F - 32) x 0.56

*It is common practice to convert from miles per gallon (mpg) to litres/100 kilometres (l/100km), where mpg (Imperial) x l/100 km = 282 and mpg (US) x l/100 km = 235

Working conditions and tools

When a major overhaul is contemplated, it is important that a clean, well-lit working space is available, equipped with a workbench and vice, and with space for laying out or storing the dismantled assemblies in an orderly manner where they are unlikely to be disturbed. The use of a good workshop will give the satisfaction of work done in comfort and without haste, where there is little chance of the machine being dismantled and reassembled in anything other than clean surroundings. Unfortunately, these ideal working conditions are not always practicable and under these latter circumstances when improvisation is called for, extra care and time will be needed.

The other essential requirement is a comprehensive set of good quality tools. Quality is of prime importance since cheap tools will prove expensive in the long run if they slip or break when in use, causing personal injury or expensive damage to the component being worked on. A good quality tool will last a long time, and more than justify the cost.

For practically all tools, a tool factor is the best source since he will have a very comprehensive range compared with the average garage or accessory shop. Having said that, accessory shops often offer excellent quality tools at discount prices, so it pays to shop around. There are plenty of tools around at reasonable prices, but always aim to purchase items which meet the relevant national safety standards. If in doubt, seek the advice of the shop proprietor or manager before making a purchase.

The basis of any tool kit is a set of open-ended spanners, which can be used on almost any part of the machine to which there is reasonable access. A set of ring spanners makes a useful addition, since they can be used on nuts that are very tight or where access is restricted. Where the cost has to be kept within reasonable bounds, a compromise can be effected with a set of combination spanners – open-ended at one end and having a ring of the same size on the other end. Socket spanners may also be considered a good investment, a basic 3/8 in or 1/2 in drive kit comprising a ratchet handle and a small number of socket heads, if money is limited. Additional sockets can be purchased, as and when they are required. Provided they are slim in profile, sockets will reach nuts or bolts that are deeply recessed. When purchasing spanners of any kind, make sure the correct size standard is purchased. Almost all machines manufactured outside the UK and the USA have metric nuts and bolts, whilst those produced in Britain have BSF or BSW sizes. The standard used in USA is AF, which is also found on some of the later British machines. Others tools that should be included in the kit are a range of crosshead screwdrivers, a pair of pliers and a hammer.

When considering the purchase of tools, it should be remembered that by carrying out the work oneself, a large proportion of the normal repair cost, made up by labour charges, will be saved. The economy made on even a minor overhaul will go a long way towards the improvement of a toolkit.

In addition to the basic tool kit, certain additional tools can prove invaluable when they are close to hand, to help speed up a multitude of repetitive jobs. For example, an impact screwdriver will ease the removal of screws that have been tightened by a similar tool, during assembly, without a risk of damaging the screw heads. And, of course, it can be used again to retighten the screws, to ensure an oil or airtight seal results. Circlip pliers have their uses too, since gear pinions, shafts and similar components are frequently retained by circlips that are not too easily displaced by a screwdriver. There are two types of circlip pliers, one for internal and one for external circlips. They may also have straight or right-angled jaws.

One of the most useful of all tools is the torque wrench, a form of spanner that can be adjusted to slip when a measured amount of force is applied to any bolt or nut. Torque wrench settings are given in almost every modern workshop or service manual, where the extent to which a complex component, such as a cylinder head, can be tightened without fear of distortion or leakage. The tightening of bearing caps is yet another example. Overtightening will stretch or even break bolts, necessitating extra work to extract the broken portions.

As may be expected, the more sophisticated the machine, the greater is the number of tools likely to be required if it is to be kept in first class condition by the home mechanic. Unfortunately there are certain jobs which cannot be accomplished successfully without the correct equipment and although there is invariably a specialist who will undertake the work for a fee, the home mechanic will have to dig more deeply in his pocket for the purchase of similar equipment if he does not wish to employ the services of others. Here a word of caution is necessary, since some of these jobs are best left to the expert. Although an electrical multimeter of the AVO type will prove helpful in tracing electrical faults, in inexperienced hands it may irrevocably damage some of the electrical components if a test current is passed through them in the wrong direction. This can apply to the synchronisation of twin or multiple carburettors too, where a certain amount of expertise is needed when setting them up with vacuum gauges. These are, however, exceptions. Some instruments, such as a strobe lamp, are virtually essential when checking the timing of a machine powered by CDI ignition system. In short, do not purchase any of these special items unless you have the experience to use them correctly.

Although this manual shows how components can be removed and replaced without the use of special service tools (unless absolutely essential), it is worthwhile giving consideration to the purchase of the more commonly used tools if the machine is regarded as a long term purchase Whilst the alternative methods suggested will remove and replace parts without risk of damage, the use of the special tools recommended and sold by the manufacturer will invariably save time.

Chapter 1 Engine, clutch and gearbox

For information relating to 1981-on models refer to Chapter 7

Contents

Specifications

Engine

Type	Four-cylinder, single overhead camshaft, air-cooled, four-stroke
Bore	59.8 mm (2.354 in)
Stroke	55.8 mm (2.197 in)
Capacity	626 cc (38.2 cu in)
Compression ratio	9.0 : 1
Claimed horsepower (max)	63 BHP @ 9000 rpm

Valve timing
Inlet opens	5° BTDC
Inlet closes	35° ABDC
Exhaust opens	40° BBDC
Exhaust closes	5° ATDC

Cylinder head
Max cylinder head warpage	0.25 mm (0.010 in)

Valves and springs
Valve stem outside diameter:	
Inlet	5.475 – 5.490 mm (0.2156 – 0.2161 in)
Service limit	5.47 mm (0.215 in)
Exhaust	5.455 – 5.470 mm (0.2148 – 0.2154 in)
Service limit	5.45 mm (0.215 in)
Valve guide internal diameter:	
Inlet/exhaust	5.500 – 5.515 mm (0.2165 – 0.2171 in)
Service limit	5.55 mm (0.219 in)
Valve stem to guide clearance:	
Inlet	0.010 – 0.040 mm (0.004 – 0.0016 in)
Service limit	0.08 mm (0.003 in)
Exhaust	0.030 – 0.050 mm (0.0012 – 0.0020 in)
Service limit	0.10 mm (0.004 in)
Valve stem runout (max)	0.05 mm (0.002 in)
Valve length:	
Inlet	90.2 mm (3.551 in)
Service limit	89.7 mm (3.531 in)
Exhaust	88.7 mm (3.492 in)
Service limit	88.0 mm (3.465 in)
Valve seat width	1.2 mm (0.047 in)
Service limit	1.5 mm (0.06 in)
Valve spring free length:	
Inner	39.2 mm (1.54 in)
Service limit	37.9 mm (1.49 in)
Outer	44.85 mm (1.77 in)
Service limit	43.3 mm (1.70 in)
Valve clearances:	
Inlet	0.05 mm (0.002 in)
Exhaust	0.08 mm (0.003 in)

Camshaft
Overall lobe height:	
Inlet	35.627 – 35.787 mm (1.4026 – 1.4089 in)
Service limit	35.6 mm (1.402 in)
Exhaust	35.314 – 35.474 mm (1.3903 – 1.3966 in)
Service limit	35.3 mm (1.390 in)
Side clearance:	
Standard	0.035 – 0.050 mm (0.0013 – 0.0020 in)
Service limit	0.1 mm (0.004 in)
Camshaft/journal clearance:	
Standard	0.160 – 0.202 mm (0.0063 – 0.0080 in)
Service limit	0.21 mm (0.008 in)

Piston and rings
Piston outside diameter:	
Standard	59.77 – 59.79 mm (2.353 – 2.354 in)
Service limit	59.65 mm (2.348 in)
Gudgeon pin bore:	
Standard	15.002 – 15.008 mm (0.5906 – 0.5909 in)
Service limit	15.08 mm (0.594 in)
Connecting rod small end internal diameter:	
Standard	15.016 – 15.034 mm (0.5912 – 0.5919 in)
Service limit	15.07 mm (0.593 in)
Gudgeon pin OD:	
Standard	14.994 – 15.000 mm (0.5903 – 0.5906 in)
Service limit	14.98 mm (0.590 in)
Gudgeon pin bore/gudgeon pin clearance (max)	0.04 mm (0.002 in)
Piston ring end gap:	
Top and 2nd ring	0.10 – 0.30 mm (0.004 – 0.0012 in)
Service limit	0.7 mm (0.028 in)
Oil control ring (side rail)	0.3 – 0.9 (0.012 – 0.035)
Service limit	1.1 mm (0.043 in)

Piston ring side clearance:
    Top and 2nd ring .... 0.015 – 0.045 mm (0.0006 – 0.0018 in)
    Service limit .... 0.15 mm (0.006 in)

Cylinder barrels

    Bore diameter .... 59.800 – 59.810 mm (2.3543 – 2.3547 in)
    Service limit .... 59.90 mm (2.358 in)
    Cylinder barrel warpage .... 0.25 mm (0.010 in)
    Cylinder/piston clearance (max) .... 0.10 mm (0.004 in)

Crankshaft and connecting rods

    Crankshaft run-out (max) .... 0.05 mm (0.002 in)
    Connecting rod big-end axial float:
        Standard .... 0.12 – 0.27 mm (0.005 – 0.011 in)
        Service limit .... 0.35 mm (0.014 in)
    Big-end bearing clearance:
        Standard .... 0.018 – 0.047 mm (0.0007 – 0.0019 in)
        Service limit .... 0.08 mm (0.003 in)
    Main bearing clearance:
        Standard .... 0.020 – 0.048 mm (0.0008 – 0.0019 in)
        Service limit .... 0.08 mm (0.003 in)
    Cam chain length .... 184.87 – 184.90 mm (7.279 – 7.280 in)
    Service limit .... 186.4 mm (7.34 in)

Clutch

    Type .... Wet, multiplate
    No of plates
        Plain 7
        Friction 8
    Plain plate warpage (max) .... 0.3 mm (0.012 in)
    Friction plate (outermost) thickness:
        Standard .... 3.42 – 3.58 mm (0.135 – 0.141 in)
        Service limit .... 3.2 mm (0.13 in)
    Friction plate (1 to 7) thickness:
        Standard .... 2.62 – 2.78 mm (0.103 – 0.109 in)
        Service limit .... 2.4 mm (0.09 in)
    No of springs 4
    Spring free length .... 36.8 mm (1.45 in)
    Service limit .... 35.4 mm (1.39 in)
    Outer drum ID .... 39.990 – 30.005 mm (1.1807 – 1.1813 in)
    Service limit .... 30.05 mm (1.183 in)

Gearbox

    Type .... 5-speed, constant mesh
    Primary reduction .... 2.737 : 1
    Final drive ratio .... 2.500 : 1 (16/40)
    Gear ratios (overall):
        1st gear .... 2.500 : 1
        2nd gear .... 1.722 : 1
        3rd gear .... 1.333 : 1
        4th gear .... 1.074 : 1
        5th gear .... 0.885 : 1

Main torque wrench settings

    Cylinder head:
        8 mm bolts .... 2.4 – 3.0 kgf m (17.22 lbf ft)
        6 mm bolts .... 1.0 – 1.4 kgf m (7 – 10 lbf ft)
    Cylinder head cover bolts (6 mm) .... 0.8 – 1.2 kgf m (6 – 9 lbf ft)
    Connecting rod nuts .... 2.4 – 2.8 kgf m (17 – 20 lbf ft)
    Cam sprocket bolts (7 mm) .... 1.4 – 1.8 kgf m (10 – 13 lbf ft)
    Crankcase bolts (8 mm) .... 2.2 – 2.6 kgf m (16 – 19 lbf ft)
    Alternator rotor centre bolt (10 mm) .... 5.0 – 6.0 kgf m (36 – 43 lbf ft)
    Clutch centre nut (20 mm) .... 4.7 – 5.3 kgf m (34 – 38 lbf ft)
    Oil filter centre bolt .... 2.7 – 3.3 kgf m (20 – 24 lbf ft)
    Engine mounting bolts:
        8 mm bolts .... 2.6 – 3.2 kgf m (19 – 23 lbf ft)
        10 mm bolts .... 3.0 – 4.0 kgf m (22 – 29 lbf ft)
        12 mm bolt .... 8.0 – 10.0 kgf m (58 – 72 lbf ft)

1 General description

The engine unit fitted to the Honda CB650 models is of the four-cylinder, air cooled type, built in unit with the gearbox, and mounted transversely in the frame. The aluminium alloy die cast crankcase is separated horizontally, with all the shafts, other than the primary shaft, lying on the jointing line. Aluminium alloy is also used for the cylinder block, which has steel insert cylinder liners, and for the cylinder head.

The crankshaft is a one-piece forging, and runs in five journals with split shell bearings. The H-section connecting rods also have shell bearings in the split big-ends, whereas the small-ends are plain bushes. The alternator is mounted directly on the left-hand end of the crankshaft and the electronic ignition unit on the right-hand end.

The pistons have full skirts, and a flat top with cutaways for the valves. There are two compression rings and a three-piece oil control ring. The gudgeon pin is fully floating.

The single overhead camshaft, driven by an endless Hy-Vo type chain from the centre of the crankshaft, runs in plain unlined bearings in the cylinder head. The chain passes through a tunnel between the centre cylinders and is tensioned by a self-adjusting rubber-faced slipper. The rockers, which bear directly on the cams, are supported on shafts in the rocker cover.

Primary drive is by a short non-adjustable Hy-Vo inverted tooth chain, from the centre of the crankshaft. It drives the primary shaft, mounted below the gearbox mainshaft and layshaft, through a rubber damper block type shock absorber assembly. The right-hand end of the primary shaft carries the straight-cut spur primary drive pinion; the left-hand end drives the oil pump. The starter gear is also mounted on the primary shaft, and drives it through a centrifugal starter clutch. The primary shaft is supported in two ball journal bearings. A spur gear on the back of the clutch outer drum meshes with the primary drive pinion, and transmits the drive to the layshaft via a multiplate wet clutch.

An Eaton trochoid oil pump is fitted. Oil is picked up from the sump via a chamber integral with the upper crankcase half, the mouth of which is closed by a detachable wire mesh screen. The screen protects the oil pump from any larger impurities which may have contaminated the oil. The oil pump forces the oil, under pressure, through a full-flow paper-element oil filter which is housed within a chamber at the front of the crankcase.

The CB650 models all employ a five-speed gearbox. Gear selection is by means of dogs on sliding gears, which are moved by three selector forks. The forks are moved in turn by a rotary selector drum. The mainshaft and layshaft each have a ball journal bearing and a needle roller bearing.

2 Operations with the engine/gearbox unit in the frame

It is not necessary to remove the engine/gearbox unit from the frame unless the crankshaft or the gearbox requires attention. Theoretically, it is possible to give the engine unit a top overhaul with the engine in the frame. Other components, such as the clutch, gear selector mechanism, alternator, etc are all readily accessible without need to remove the engine unit. Tasks that can be carried out with the engine in situ are as follows:

(a) *Removal of cylinder head, cylinder block and pistons*
(b) *Removal of the clutch*
(c) *Removal of the alternator and starter motor*
(d) *Removal of the carburettors*
(e) *Removal of the gearchange external selector mechanism*
(f) *Removal of the oil pump and oil filter*

When several tasks have to be undertaken simultaneously, it will probably prove advantageous in terms of better access and more working space, to remove the complete engine unit from the frame.

Note that the crankcase cannot be separated until the engine is removed from the frame and that access is not available to the gearbox until this has been accomplished. It is, however, possible to overhaul the gearbox, after separation, without need to disturb the cylinders. The crankcase must be rejoined before it can be replaced in the frame.

3 Operations with the engine/gearbox unit removed from the frame

(a) *Removal and replacement of the main bearings and big-ends*
(b) *Removal and replacement of the crankshaft assembly*
(c) *Removal and replacement of the Hy-Vo primary chain*
(d) *Removal and replacement of the gearbox components including the gearchange internal selection mechanism*

4 Method of engine/gearbox unit removal

As described previously, the engine and gearbox are built in unit and it is necessary to remove the complete assembly in order to gain access to either sub-assembly. Separation of the crankcase is achieved after the engine unit has been removed from the frame and refitting cannot take place until the engine/gear unit is assembled completely. Access to the gearbox is not available until the engine has been dismantled and vice versa in the case of attention to the bottom end of the engine. Fortunately, the task is made easy by arranging the crankcases to separate horizontally.

5 Removing the engine/gearbox from the frame

1 Place the machine on the centre stand, so that it is standing firmly on level ground. Place a receptacle that will hold at least a gallon (4 litres) under the crankcase and remove the drain plug so that the oil will drain off. It is preferable to do this whilst the engine is warm, so that the oil will drain more readily.
2 Place a second smaller container below the oil filter chamber at the front of the engine. Unscrew the central bolt and remove the filter and its housing. Take care, because the filter housing will be full of oil. Leave the oil to drain whilst continuing the dismantling procedure.
3 On CB650 C and CA models detach the dualseat, by removing the two mounting bolts, to gain access to the petrol tank rear securing bolt. On other models merely raise the dualseat on its hinge. Turn the fuel tap lever to the OFF position and disconnect from the tap unions, the petrol feed pipe. The fuel pipe is secured by spring clips, the ears of which should be squeezed together to release their grip on the pipe. Remove the single bolt which passes through the lug at the rear of the tank and into the frame, and also remove the rubber support block. The tank is supported at the front by steel cups which locate with a rubber buffer each side of the frame top tube. Drainage of the tank is not necessary, although to do so will reduce the overall weight and make removal easier. A full tank will weigh in the region of 30 – 40 lbs. Ease the tank rearwards until the cups leave the rubbers and then lift the tank away from the machine.
4 Pull off both side covers. On the right-hand side, disconnect both battery leads from the terminals. Isolating the electrical system in this way will prevent accidental shorting of wires which are subsequently disconnected. Disconnect the starter motor cable at the block connector on the starter relay. Similarly, disconnect the lead for the electric ignition system at the block connector behind the left-hand side panel.
5 Loosen the screw clips which secure the carburettor mouths to the air cleaner forward chamber intake hoses. Loosen the large screw clip that secures the rear of the forward chamber to the front of the air filter element-containing case. Disconnect the large diameter breather hose at the breather cover on the cylinder head. Push the air cleaner chamber to the rear to separate the intake hoses from the carburettor mouths.

Loosen the screw clips which secure the carburettors to the inlet stubs. It should now be possible to remove the bank of carburettors. Because of the limited space in this area, and the substantial size of the air cleaner chamber, some careful manoeuvring will almost certainly be necessary to allow the carburettors, and then the filter chamber, to be pulled clear.

6   As the carburettors are pulled free, disconnect both the throttle cables and the choke cable. The carburettors will pull free with the four drain tubes attached. Note that they reside to the right-hand side of the rear of the engine. If desired, on the CB650 A and CA models, the two chromed air cleaner chamber end covers may be detached. Each cover is retained by two bolts. Remove the chamber and the breather tube fitted to the left-hand end of the chamber.

7   Pull off the sparking plug HT leads, they are numbered 1 to 4 from left to right to aid correct refitting. Rest the leads over the frame main top member.

8   The tachometer drive is adjacent to the right-hand rocker cover. Remove the securing bolt and extract the drive cable. Replace the bolt to avoid loss.

9   Disconnect the clutch cable from the operating arm below the clutch cover on the right-hand side of the engine. This can be accomplished most easily by slackening the clutch cable adjusters and then operating the clutch actuating arm with a suitable spanner so that the cable nipple can be displaced.

10   Separate the left-hand block connector of the four connectors situated above the air filter case (CB650 Z models) or the upper of the two connectors behind the left-hand side panel (CB650 A and CA models). The connector contains five wires (3 yellow, 1 white and 1 black) which emanate from the alternator. Similarly, disconnect the smaller connector on the left-hand side of the machine, adjacent to the rear of the top frame tube, containing the wires for the neutral indicator lamp (light green/red tracer) and the oil pressure warning lamp (blue/red tracer), and the connector on the right-hand side of the machine for the pulser generator leads (2 yellow, 2 blue).

11   Slacken evenly and remove the two nuts holding each finned exhaust port flange. Allow the flanges to slide down the exhaust pipes, catching the two split collars on each down-pipe as they fall free. On all twin silencer models, the outer exhaust down-pipes (Nos 1 and 4) are integral with the silencers, and the inner down-pipes are separate. On the four silencer set-up as fitted to the 'Custom' model, each silencer is integral with its down-pipe. The pipes from cylinders No 1 and 2 and cylinders No 3 and 4 are paired together by means of a small balance chamber interconnecting the lower ends of each pair of down-pipes. Each silencer (or each pair of silencers) is retained to the rear of the machine by a single bolt. On the CB650 Z model, the silencer retaining bolts pass through the rear of the footrest mounting plates. Remove the silencer mounting bolts, pull the down-pipes forward slightly to clear the exhaust ports, whilst supporting the silencers (an assistant is useful at this juncture) and manoeuvre the complete exhaust system clear. On the CB650 Z model, the down-pipes for cylinders No 1 and 2 will lift away together complete with the left-hand silencer. This operation will be facilitated by lowering the side stand. The down-pipes for the other two cylinders (Nos 3 and 4) and the right-hand silencer can similarly be removed as an assembly. On the CB650 A and CA models, the two pairs of down-pipes and silencers can be removed in a similar fashion.

12   The engine oil should now have fully drained. Remove the receptacle containing the drained oil and continue with the dismantling procedure.

13   Remove the gearchange lever. On US models this is simply accomplished by removing the pinch bolt and drawing the lever off the splined shaft. Refit the bolt to avoid loss. Note the position of the lever for refitting. On the CB650 Z (UK) model, a short rear-set linkage between the gearchange lever and selector shaft is employed. On this model, the linkage is secured by a pinch bolt to the selector shaft splines, whilst a circlip retains the lever to its pivot on the left-hand footrest mounting plate. Displace the lever circlip and remove the washer. Remove the linkage pinch bolt and pull the lever and linkage off together.

Note the punch mark on the linkage and splined shaft, to aid correct replacement.

14   Remove the four retaining bolts and detach the left-hand rear crankcase cover. Note the position of the two dowels. There is no gasket used on this cover.

15   Because the chain is of the endless type (ie has no spring link) the chain and the drive sprocket must be removed simultaneously from the gearbox output shaft. Separating the chain with a rivet extractor should not be attempted unless the chain is to be discarded. If the chain was separated in this way, the resultant weak point would be a major safety hazard; the chain separating under high stress conditions might be both extremely dangerous and very costly in subsequent mechanical repairs. A certain amount of slack will be necessary to permit removal of the gearbox sprocket. Withdraw the split pin which secures the rear wheel spindle nut. Slacken the nut, and release the chain adjusters so that the wheel can be pushed forwards.

16   Instruct an assistant to apply the rear brake of the machine whilst the two sprocket retaining plate bolts are loosened and removed. Remove the retaining plate by turning it until the splines can be engaged. Pull the sprocket off the splined shaft, still meshed with the chain. Disengage the sprocket from the chain, and allow the latter to rest on the swinging arm cross member, whilst the sprocket is placed to one side.

17   From the right-hand side of the machine, detach the alloy footrest/brake pedal mounting plate (CB650 Z UK model) or the riders' right-hand footrest and the rear brake pedal (all other models). The right-hand mounting plate on the UK market CB650 Z model is secured by the swinging arm pivot shaft nut and the rear engine mounting bolt nut. Remove both nuts and then detach the brake operating rod from the top of the operating arm at the rear wheel. Also disconnect the rear brake light switch operating spring. The complete mounting plate can now be lifted clear of the machine. On the other models remove the front footrests and rear brake pedal, these being attached in a similar manner to that described above, but without the alloy mounting plate. On these models, centre punch the brake pedal and shaft if not already marked to aid replacement in the same position. Remove the brake pedal pinch bolt and pull the pedal from the splined shaft. Replace the bolt.

18   Remove the nut on the right-hand end of the engine top rear mounting bolt and remove the battery earth lead terminal.

19   The engine/gearbox unit of these models, although not being particularly heavy, is somewhat cumbersome, and as such is not easily manhandled. It is recommended, therefore, that at least two people are present when lifting the engine from place. This alleviates the risk of damage to both the engine and the operator. Before removing any of the engine mounting bolts, the acquisition of a jack will be very beneficial. By positioning the jack below the sump, much of the weight can be taken when the mounting bolts are removed. Protect the delicate sump fins with rags or a flat wooden board. A somewhat unusual and useful feature of the frame on these models, is the detachable lower right-hand frame rail which facilitates engine removal.

20   Undo and remove the two bolts at the rear of the lower right-hand frame rail, followed by the single lower forward bolt, and then the three upper bolts on the right-hand front cradle downtube. Detach the lower right-hand frame rail. Remove the lower single left-hand mounting bolt, followed by the three bolts from the front left-hand cradle down-tube. Remove the remaining two upper rear bolts on the right-hand side of the frame. The engine is now only retained by the two long rear mounting bolts. With the jack in position, or alternatively at least one assistant to support the engine, withdraw the lower rear bolt and then the upper rear bolt. Note that there is a spacer between the frame and the crankcase on the left-hand end of the upper rear bolt.

21   After removal of all the bolts, the engine will have settled onto the jack allowing the operator, and his assistant, time to find a handhold. Lift the engine up gradually and out of the frame from the right-hand side. Place the engine on a clean flat surface to await further attention.

5.5a Slacken the screw clips on the air cleaner chamber intake hoses and ...

5.5b ... the forward intake hose screw clips

5.8 Remove single retaining bolt to free tachometer drive cable

5.11a Exhaust pipes are secured by two nuts and finned flange; note the split collars

5.11b Silencer is retained to mounting plate by single bolt (CB650Z shown)

5.13 Remove short rearset linkage that operates gearchange (UK CB650Z)

5.16 Release bolts and retainer plate to withdraw sprocket still meshed with chain

5.17 Remove footrests/brake pedal mounting plate (UK CB650Z)

5.19 Engine/gearbox unit is now ready for removal; note jack below sump

6 Dismantling the engine/gearbox unit: general

1   Before commencing work on the engine, the external surfaces must be cleaned thoroughly. A motorcycle engine has very little protection from road dirt, which will sooner or later find its way into the dismantled engine if this simple precaution is not observed.
2   One of the proprietary engine cleaning compounds such as Gunk or Jizer can be used to good effect, especially if the compound is allowed to penetrate the film of oil and grease before it is washed away. When washing down, make sure that water cannot enter the carburettors or the electrical system, particularly if these parts are now more exposed.
3   When clean and dry, arrange the unit on the workbench, leaving a suitable clear area for working. Gather together a selection of containers for small items (plastic tubs and egg cartons are useful) which can then be easily kept grouped together. Some paper and a pen should be on hand to permit notes to be made and labels attached where necessary. A supply of clean rag is also required.
4   An assortment of tools will be required, in addition to those supplied with the machine (see Working conditions and Tools for details). Unlike owners of most Japanese machines, those

working on the CB650 are fortunate in not having large numbers of the normal soft alloy, crosshead screws to cope with. In most cases, small hexagon-headed bolts are employed, these being much more robust. In view of this, an appropriate 'nut driver' or box spanner will prove invaluable, in addition to the normal range of workshop tools.
5   Before commencing work, read through the appropriate sections so that some idea of the necessary procedure can be gained. Never use force to remove any stubborn part, unless mention is made of this requirement in the text. There is invariably good reason why a part is difficult to remove, often because the dismantling operation has been tackled in the wrong sequence.

7 Dismantling the engine/gearbox unit: removing the rocker gear and camshaft

1   Remove the individual tappet adjusting covers from the outer cylinders; each cover is held by two bolts. Remove the central breather cover, from above the inner two cylinders. The breather cover is retained by two bolts. Each of the eight cover bolts is fitted with a rubber seal, and the covers themselves are equipped with rubber seals.
2   Remove the four sparking plugs. The correct plug spanner (as supplied in the original tool kit) should be used to facilitate plug removal. Due to the small size and recessed fitting of the plugs, an alternative tool will prove difficult to find.
3   Loosen and remove the cylinder head cover retaining bolts (22 in all). These bolts should be slackened in two or three stages in a uniform and diagonal manner. This will relieve stress gradually and evenly, therefore negating the chance of distortion. Note the fitting of two dowels to the rear of the cylinder head cover. The cover may now be put to one side for further dismantling and inspection at a later stage.
4   Remove the two retaining bolts and remove the circular CDI pulser assembly cover, from the right-hand side of the engine. Using a spanner on the large hexagonal spacer, fitted outboard of the pulser assembly, turn the crankshaft forwards (clockwise) until one of the cam sprocket bolts comes to the top. Remove this bolt. Rotate the crankshaft 180°, until the second bolt can be removed. Ensure the bolts are not dropped into the depths of the crankcase.
5   To facilitate camshaft removal some slack in the cam chain is desirable. Slacken the cam chain adjuster locknut at the rear of the cylinder block and lift the camshaft so that the rear run

of the cam chain is straightened. This will push back the tensioner blade which can then be locked in position by tightening the locknut.

6 Slide the sprocket to the left, off its locating shoulder and by careful 'wriggling', manoeuvre the camshaft nut to the right of the sprocket. The cam sprocket and chain can be held by hand during camshaft removal. With the camshaft removed, disengage the sprocket from the cam chain. Hook a stout piece of wire, or a screwdriver, through the chain to prevent it dropping into the tunnel.

7 Remove the cam chain tensioner blade locating bolt at the rear top of the cylinder head, above the adjuster.

8 Dismantling the engine/gearbox unit: removing the cylinder head

1 The cylinder head is retained by twelve 8 mm bolts and an additional two 6 mm bolts. The two 6 mm bolts are fitted to the front and rear of the cam chain tunnel. Remove the six sealing rubbers from the inner bolts. Slacken and remove the cylinder head bolts in the reverse order to that shown in the bolt tightening diagram in Section 50. The bolts must be loosened in this sequence, in two or three stages, to avoid stressing the head casting unevenly.

2 Separate the cylinder head from the very tenacious gasket using a soft-headed mallet. Strike only those portions of the casting which are adequately supported, taking special care not to damage the fins. Under no circumstances should levers be used between the mating surfaces of the cylinder head and cylinder block in an effort to facilitate separation. This action will lead to damage to the faces with subsequent risk of leakage.

3 Lift off the cylinder head, guiding the cam chain through the central tunnel, and preventing it from falling free. An extra pair of hands is beneficial at this stage. Put the head down on a clean flat surface. When the cylinder head has been removed, replace the cam chain support on the top of the cylinder block.

4 Note the two dowels at the rear of the cylinder block and the two O-rings at each end. Remove these and store them safely until reassembly.

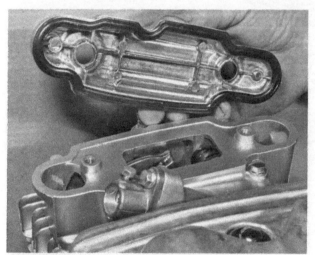

7.1a Remove the outer cylinders' tappet inspection covers and ...

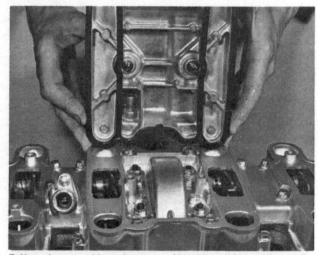

7.1b ... the central breather cover. Note the rubber seals on all covers

7.3 Release and remove cylinder head cover

7.7 Remove the cam chain tensioner bolt

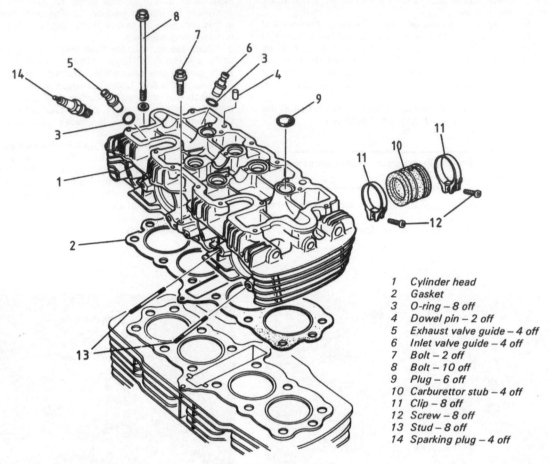

1   Cylinder head
2   Gasket
3   O-ring – 8 off
4   Dowel pin – 2 off
5   Exhaust valve guide – 4 off
6   Inlet valve guide – 4 off
7   Bolt – 2 off
8   Bolt – 10 off
9   Plug – 6 off
10   Carburettor stub – 4 off
11   Clip – 8 off
12   Screw – 8 off
13   Stud – 8 off
14   Sparking plug – 4 off

Fig. 1.1 Cylinder head

8.1 Remove the 6 mm bolts to the front and rear of the cylinder head

9   Dismantling the engine/gearbox unit: removing the cylinder block

1   Remove the cam chain guide from the front of the cylinder block by raising the guide slightly, rotating it 90° and withdrawing it. Take care not to drop the cam chain into the crankcase during this operation. Unscrew the cam chain adjuster locknut and remove the tensioner unit from the cylinder block by pulling the unit upwards. If it is found difficult to remove the unit it can be easily released from the cylinder block after the block has been removed.
2   Pull up the cam chain and turn the crankshaft until all the pistons are at approximately the same position in their bores.
3   Separate the cylinder block from the 'sticky' base gasket using the technique described for the cylinder head. Once again, levers should not be used. Lift the cylinder block upwards along the holding down studs until the piston skirts are visible but the piston rings are still obscured by the cylinder bore spigots. If a top-end overhaul only is to be carried out, the crankcase mouths must be padded with clean rags to prevent any pieces of broken piston ring from falling into the crankcase. The padding will also prevent the ingress of foreign matter during further dismantling or work. With the padding in place, lift the cylinder block off the pistons as squarely as possible to prevent the pistons from tying in the bore. Catch the pistons and connecting rods as they emerge from their bores (a second pair of hands is useful here) to prevent damage occurring.
6   Once clear of all four pistons, the cylinder block can be placed to one side for further examination later.
7   Note the two cylinder block locating dowels, complete with O-ring seals, which double as oil metering jets.

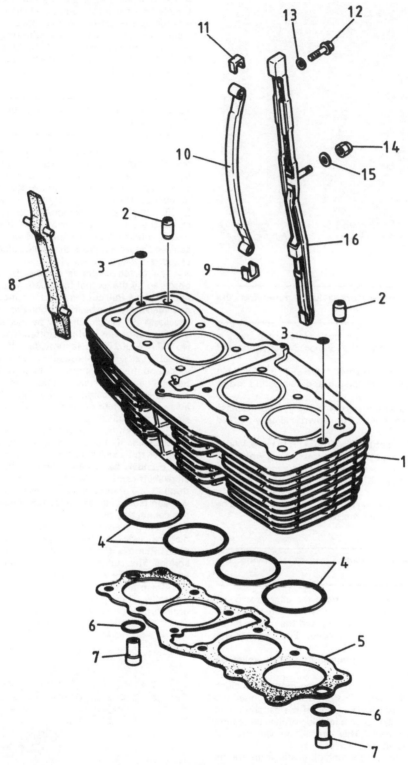

Fig. 1.2 Cylinder block and cam chain tensioner

1	Cylinder block	7	Hollow dowel – 2 off	12	Bolt
2	Hollow dowel – 2 off	8	Chain guide blade	13	Washer
3	O-ring – 2 off	9	Blade lower seat	14	Domed nut
4	O-ring – 4 off	10	Tensioner blade	15	Washer
5	Base gasket	11	Blade upper seat	16	Chain tensioner unit
6	O-ring – 2 off				

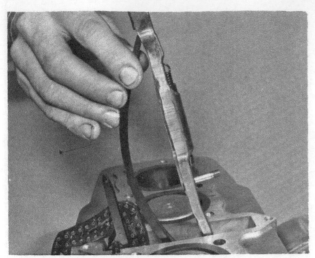

9.1 Withdraw the tensioner unit

10 Dismantling the engine/gearbox unit: removing the pistons and piston rings

1   Remove both wire circlips from each piston boss using a small screwdriver blade. Discard the circlips. Never re-use old circlips during the rebuild.
2   Using a drift of the correct diameter, tap each gudgeon pin out of position, supporting each piston and connecting rod in turn.
3   Mark each piston inside the skirt with the cylinder number (1-4, left to right), as it is removed. The crown is already marked to indicate the rear.
4   If the gudgeon pin is a tight fit **do not** resort to force. Warm the piston by wrapping around the piston crown a rag which has been soaked in very hot water, and then wrung out. The resultant expansion should ease the grip of the piston bosses on the steel pins.
5   Do not remove the piston rings at this stage; they should be left in place on the pistons until the examination stage.

11 Dismantling the engine/gearbox unit: removing the alternator

1   Remove the four retaining bolts from the alternator cover on the left-hand side of the engine. The cover can then be lifted away complete with the stator assembly and the pickup brush holder. Note the two dowels on the cover. Ensure the cover gasket is not damaged.
2   It will be necessary to prevent crankshaft rotation whilst the rotor securing bolt is removed.
3   If the cylinder head and block have been removed, a close fitting bar can be passed through one or two of the connecting rod eyes to lock the crankshaft. The ends of the bar should be rested on wooden blocks placed on each side of the connecting rod on the crankcase mouth. This precaution will prevent damage to the crankcase mating surface.
4   An alternative procedure, if the engine is still in the frame, is to select top gear and apply the rear brake to lock the entire power train.
5   After the rotor securing bolt has been removed, it will be necessary to contrive some means of drawing the rotor off the crankshaft taper. The rotor boss is fitted with a large internal thread which is designed to accept a Honda service tool – rotor puller (part number 07733-0020001). If this tool is not available, a suitable alternative can be found in the form of the rear wheel spindle. This has the correct thread and a taper

which matches that of the crankshaft end. Screw the puller or spindle into the rotor centre and then tighten it down until a reasonable amount of force is applied. If this fails to break the tapered joint, do not continue tightening or damage to the rotor or crankshaft end may result. Strike the head of the spindle firmly; this should have the effect of releasing the rotor. Do not strike the rotor itself because this will almost certainly cause damage.

12 Dismantling the engine/gearbox unit: removing the clutch assembly

1   The clutch assembly can be removed and replaced irrespective of whether the engine has been removed, or is still in the frame. Where the unit is installed in the frame, disconnect the clutch operating cable, remove the rear brake pedal (and footrest on US models) and drain the engine/transmission oil.
2   Remove the ten hexagon-headed retaining screws from the clutch cover and lift the cover away. If the cover is stuck to the gasket, tap lightly around the joint with a soft-headed mallet to achieve separation. Note the cable guide clips attached to three of the retaining screws.
3   The clutch centre is secured by a central nut which is obscured by the journal ball bearing in the centre of the clutch release plate. Pull out the clutch pushrod piece from the bearing and then ease the bearing from position in the plate so that access to the clutch nut may be made.
4   The centre nut is of the slotted type for which a Honda tool, No 07716-0020202 is provided. If the proper tool is not available a suitable substitute can be constructed in the workshop from a length of thick-walled tubing. Clamp the tubing in a vice and use a hacksaw to cut slots as shown in the accompanying illustration. The shaded area can then be filed away to leave four projecting lugs which will engage with the slots in the nut.
5   The clutch must be prevented from rotating whilst the nut is loosened. If the engine is in the frame this can be accomplished by selecting top gear and applying the rear brake. If the cylinder barrel and pistons have been removed the crankshaft can be locked by passing a smooth, close-fitting bar through one or more small-end eyes, arranged so that it rests on wooden blocks placed across the crankcase mouth each side of the connecting rod.
6   Having slackened the clutch centre nut, slacken in a diagonal sequence the four bolts which retain the clutch release plate. Remove the bolts followed by the plate and pressure springs. The clutch centre nut can now be run off the thread, and the Belville washer fitted below removed.
7   Grasp the outer clutch drum and slide the complete clutch assembly off the gearbox shaft. Leave the assembly undisturbed until further examination is required. Refer to Section 33.

10.1 Prise out the piston circlips and ...

10.2 ... push out the gudgeon pin to free the piston. Mark each piston on removal

11.5 Some form of puller is necessary for rotor removal; rear wheel spindle shown inserted

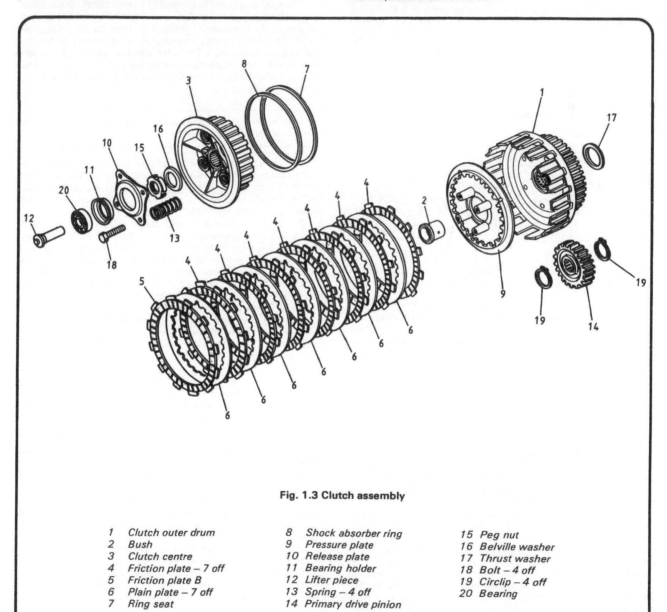

Fig. 1.3 Clutch assembly

1 Clutch outer drum	8 Shock absorber ring	15 Peg nut
2 Bush	9 Pressure plate	16 Belville washer
3 Clutch centre	10 Release plate	17 Thrust washer
4 Friction plate – 7 off	11 Bearing holder	18 Bolt – 4 off
5 Friction plate B	12 Lifter piece	19 Circlip – 4 off
6 Plain plate – 7 off	13 Spring – 4 off	20 Bearing
7 Ring seat	14 Primary drive pinion	

12.4 Remove the castellated clutch centre nut with home-made peg spanner

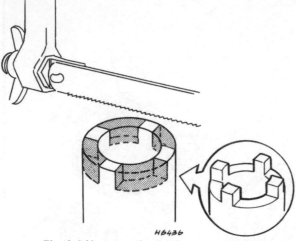

H6436

Fig. 1.4 Home-made clutch nut peg spanner

13 Dismantling the engine/gearbox unit: removing the starter motor

1   The starter motor rests in a well formed in the crankcase, immediately to the rear of the cylinder block. The motor can be removed with the engine unit installed in the frame or on the workbench. It is not essential to remove the motor to permit crankcase separation but it is usual to do so during a full overhaul. If the motor is to be removed with the engine in the frame, several operations should be carried out prior to motor removal.
2   Check that the ignition is switched off, then isolate the battery by disconnecting the negative (-) lead. Remove the right-hand side cover, and trace the heavy starter motor cable back to its terminal on the starter relay to the rear of the battery. Disconnect the cable and pull it through to clear the frame.
3   The starter motor is located beneath a chromed cover to the left of the rear of the cylinder block. The cover is retained by two bolts. Remove the cover to expose the motor and its two retaining bolts. Remove the bolts, then grasp the motor body and pull it to the left-hand side to free it from the engine. The motor is sealed by a large O-ring where it enters the crankcases and it may be necessary to lever the motor back against the O-rings' resistance. This must be done very carefully to avoid damage. Once freed, the motor can be lifted away, together with its feed cable.

14 Dismantling the engine/gearbox unit: removing the CDI pulser assembly and automatic timing unit (ATU)

1   The CDI pulser assembly and automatic timing unit are housed behind the small right-hand engine outer cover. The various components can be removed with the engine in position in or removed from the frame.
2   Remove the two hexagon-headed retaining bolts and detach the circular cover. If the pulser assembly is to be removed with the engine in position, separate the pulser lead at its connector behind the right-hand side cover. Also separate the lead from the three guide clips around the clutch cover.
3   The CDI pulser assembly is secured to the engine by three crosshead screws passing through the periphery of the baseplate. Before disturbing the plate it is well to scribe-mark the relative position of the plate against the casing so that the ignition timing can be reset easily on reassembly. Remove the three screws and detach the assembly. Note the rubber sealing grommet in the lower casing wall.
4   The ATU is secured to the end of a small diameter advancer unit shaft by a 6 mm nut. The advancer shaft itself is fitted to the end of the crankshaft. Using a spanner on the hexagonal spacer provided, prevent the crankshaft from rotating, whilst the retaining nut is removed. The ATU assembly can now be pulled clear of the shaft. Note the locating pin which engages in the back of the ATU. Check the fit of the pin, removing it if loose, to avoid accidental loss.

15 Dismantling the engine/gearbox unit: removing the gear selector external components

1   To gain access to the external gear selector components it is not necessary to remove the engine from the frame. If the operation is to be carried out with the engine in situ, first drain the engine oil, remove the right-hand crankcase cover, and remove the clutch assembly.
2   Remove the pivot bolt and collar for the selector drum detent lever and the securing bolt for the neutral detent lever arm. Note that the lower end of the neutral detent lever arm is just above the primary drive pinion, and the lower bolt also secures the top of the primary shaft bearing retainer plate.
3   Depress the selector shaft claw lever and disengage it from the end of the selector drum. Withdraw the complete selector shaft from the lower crankcase.
4   Remove the two bolts and remove the positive stop lever/selector drum bearing retainer plate. Remove the bolt which retains the star-shaped drum stopper index plate to the outer end of the gear change drum. Remove the index plate, followed by the drum end centre piece complete with four selector pins. As the centre piece is removed, note the further pin fitted in the drum end.
5   Note the location of all these components and the position of levers and their springs for reassembly.

16 Dismantling the engine/gearbox unit: removing the oil pump

1   The oil pump may be removed with the engine in the frame or on the workbench. If the pump is to be removed whilst the engine is in situ, drain the engine oil, remove the gearchange lever or lever assembly and detach the left-hand rear crankcase cover, prior to pump removal.
2   Disconnect the oil pressure warning light lead from the top of the oil pump pressure switch.
3   Unscrew the three crosshead screws which secure the pump to the crankcase and remove the pump. Note the two dowels with O-ring seals, and the single large O-ring fitted to the rear of the pump. Also note the oil control valve and its separate O-ring.

14.3 Remove the three screws and remove the pulser assembly

14.4 Withdraw the ATU from the advancer shaft; note the locating pin

17 Dismantling the engine/gearbox unit: removing the primary shaft and separating the crankcase halves

1   Commence crankcase operation with the engine sitting upright in the workbench. Remove the nine various length bolts which pass down from the upper crankcase half. These are all located around the rear section of the crankcase, behind the cylinder block gasket face. Slacken the bolts in a diagonal sequence, turning each one by about $\frac{1}{2}$ turn until all pressure has been released. This will ensure that uneven pressure is not placed on any one area, and will thus prevent any risk of warpage.

2   Turn the unit upside down, supporting it with blocks at the rear to keep it level. The unit can safely be supported on the holding studs at the front as long as they remain vertical and are not subjected to any lateral strain.

3   Remove the sump after removing its ten retaining bolts. Remove the rubber sump gasket carefully and keep it safely. Pull off the oil strainer, which is a push fit in a central bore on the underside of the lower crankcase half. Note the two O-rings on the oil strainer spigot.

4   The crankcase halves are now secured by a total of twenty six bolts. Release the bolts in gradual stages working in a criss-cross pattern to avoid any risk of warpage. Note the position of the two cable guide clips. Note also the single bolt fitted within the confines of the sump.

5   In order that the crankcase halves may be completely separated, the primary shaft must be removed. Remove the primary drive pinion from the right-hand end of the shaft. The pinion is retained by a circlip, and thrust washer. Remove the primary shaft right-hand bearing retainer plate. The retainer plate is now only secured by a single bolt. Release the primary drive chain tensioner spring by unhooking one end. The primary shaft will now require pulling forcibly from position. The recommended Honda service tool for this operation is a sliding hammer device. Part Nos 07936 – 3740100 (sliding hammer shaft) and 07945 – 30000500 (sliding hammer weight). This device consists of a headed shaft upon which is placed a free sliding steel weight. After screwing the shaft into the primary shaft, the steel weight is slid forcibly along the shaft away from the primary shaft until it contacts the shaft head. The force imparted will withdraw the primary shaft. If the correct sliding hammer is not available, an alternative tool of the same type may be simply constructed using the long lower rear engine mounting bolt and a suitable sliding weight. A large socket spanner makes an ideal weight.

6   With all the crankcase bolts removed and the primary shaft and bearing removed, final separation of the crankcase halves is possible.

8   Lift away the lower crankcase half. In this way all the shafts will remain in the upper crankcase half. The jointing compound used on the mating surfaces may tend to make separation difficult. If this is the case, tap around the joint with a soft-headed mallet to help break the seal. If necessary, use a hammer and a hardwood block against the more substantial parts of the casing, and carefully knock the lower crankcase upwards. **Do not** use levers between the mating surfaces of the crankcase halves. This will lead to damage of the surfaces and subsequent oil leakage.

18 Dismantling the engine/gearbox unit: removing the crankshaft and primary shaft components

1   Having removed the lower crankcase half, it will be noted that all the internal components have remained in the inverted upper casing half. The crankshaft, gearbox mainshaft and layshaft, the internal gear selection mechanism (selector drum, forks and support shaft), and the primary chain and primary shaft driven sprocket/starter motor clutch assembly, are all retained in the upper casing.

2   Lift up the primary shaft driven sprocket/starter motor clutch assembly and unhook the primary chain. Take care when handling the sprocket/starter clutch assembly that it is not inadvertently dropped or banged. This may cause the clutch rollers to drop out of position, necessitating the otherwise unnecessary chore of re-installing them.

3   Grasp the crankshaft with both hands and lift it upwards, out of the casing, as a complete unit together with the oil seals and both chains. If the crankshaft is firmly seated and resists removal, tap it gently with a soft-headed mallet to free it from the casing.

4   The main bearing shells will remain in the crankcase. **Do not** remove them unless they are to be renewed.

5   The primary shaft components also should now be removed from the casing. Remove the driven sprocket/starter clutch assembly, followed by the large diameter (57T) starter clutch pinion and its needle roller bearing and spacer.

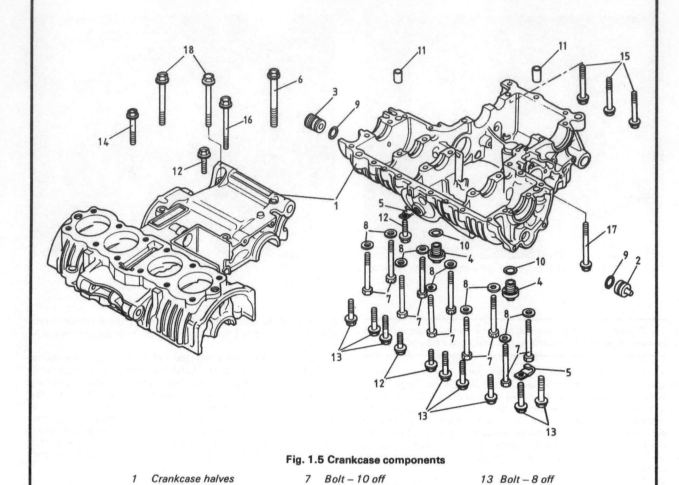

Fig. 1.5 Crankcase components

1	Crankcase halves	7	Bolt – 10 off	13	Bolt – 8 off
2	Oilway plug	8	Washer – 10 off	14	Bolt
3	Oilway plug	9	O-ring – 2 off	15	Bolt – 3 off
4	Oilway plug – 2 off	10	O-ring – 2 off	16	Bolt
5	Cable clip – 2 off	11	Hollow dowel – 2 off	17	Bolt
6	Bolt	12	Bolt – 4 off	18	Bolt – 2 off

17.3 Lift away the sump; note the rubber gasket

17.5a Remove the primary drive pinion and ...

17.5b ... pull the primary shaft out of the casing; engine mounting bolt and socket spanner will serve as extractor

17.6 Ensure primary chain tensioner spring is released when withdrawing primary shaft

19 Dismantling the engine/gearbox unit: removing the gear shafts and internal gear selector mechanism

1   If attention is required to the selector drum, the selector forks or their support shaft, the gearbox main shaft and layshaft must be removed. If the gear pinions mounted on these shafts are suspected of being excessively worn, the backlash between each pair of gears should be measured at this stage, prior to removal. This measurement requires the use of a dial gauge (DTI) mounted on a suitable stand. Arrange the probe of the dial gauge to rest upon one of the gear teeth, then set the gauge at zero. Rock the pinion back and forth and note the extent of free play (backlash). This should not exceed 0.20 mm (0.008 in) on any pair of pinions. Excessive clearance is indicative of wear, and may require renewal of the pinions concerned.
2   Lift the mainshaft and layshaft assemblies clear of the casing, complete with pinions and bearings. Note the layshaft, left-hand oil seal, it is secured by locking compound. The two shafts can now be placed to one side, in their normal operating position, until further attention or reassembly is due. Note the fitting and positioning of the bearing locating dowels and bearing locating half-rings.
3   Remove the neutral indicator switch cam after unscrewing the crosshead retaining screw in the centre of the cam. If the two bolts which secure the selector drum bearing retainer plate were not removed previously, remove the bolts and the plate now. In order that the selector drum may be removed, the drum mounted mainshaft selector fork must first be removed. The selector fork is retained by a spring clip and guide pin. Remove the spring clip with the aid of a pair of snipe-nose pliers. Remove the guide pin in a similar fashion or with the use of a small magnet (eg a magnetic screwdriver). Hold the selector fork, and withdraw the selector drum and its bearing, which is a light push fit in the casing bore.
4   The two layshaft selector forks can be released by withdrawing the support shaft from the casing. Hold the forks as the shaft is withdrawn. Note the position of each fork prior to removal, and then refit them to the support shaft in the correct manner; this will ease reassembly.
5   Remove one further crosshead screw and release the neutral indicator switch unit.
6   Remove the starter idler gear, and from the lower casing half, remove the primary chain tensioner device. Bend flat the locking tab on the locking plate securing the primary chain tensioner retaining bolt. Unscrew and remove the bolt, and lift

away the tensioner. Similarly, flatten the locking tab on the locking plate securing the bolt retaining the idler gear shaft. Unscrew the bolt. Insert a suitable long bolt into the threaded outer end of the shaft, through the hole in the casing, and withdraw the shaft. Lift away the idler gear.
7   If required, the primary chain oil feed nozzle can be removed from the lower casing half. It is situated just forward of the primary shaft left-hand ball journal bearing. The nozzle and its integral fixing plate are secured in the casing by a single self-tapping crosshead screw. Remove the screw and lift the nozzle and plate away. Note the O-ring fitted below the plate. Although removal is not essential, the oil feed nozzle can be checked for clogging whilst it is detached from the casing. Clogging of the nozzle could lead to inadequate lubrication of the primary chain and resulting damage to the whole primary drive train.
8   Check over both bore crankcase halves, and remove any dowel pins or half-rings which may have been left behind as the various shafts were removed. It is advisable to mark the various small parts to avoid confusion during reassembly.

19.3 Remove neutral switch cam retaining screw (arrowed)

19.7 Primary chain oil feed nozzle may be removed for cleaning

20 Examination and renovation: general

1   Before examining the parts of the dismantled engine unit for wear, it is essential that they should be cleaned thoroughly. Use a paraffin/petrol mix to remove all traces of old oil and sludge that may have accumulated within the engine.
2   Examine the crankcase castings for cracks or other signs of damage. If a crack is discovered, it will require professional repair, or renewal.
3   Carefully examine each part to determine the extent of wear, checking with the tolerance figures listed in the text or in the Specifications Section of this Chapter. If there is any question of doubt, play safe and renew.
4   Use a clean lint free rag for cleaning and drying the components. This will obviate the risk of small particles obstructing the internal oilways, causing the lubrication system to fail.

21 Big-end and main bearings: examination and renovation

1   The Honda CB650 models, in common with most other Honda motorcycles, are fitted with renewable shell-type plain bearings on the crankshaft main and big-end journals.
2   Bearing shells are relatively inexpensive and it is prudent to renew the entire set of main bearing shells when the engine is dismantled completely, especially in view of the amount of work which will be necessary at a later date if any of the bearings fail. Always renew the five sets of main bearings together.
3   Wear is usually evident in the form of scuffing or score marks in the bearing surface. It is not possible to polish these marks out in view of the very soft nature of the bearing surface, and the increased clearance that will result. If wear of this nature is detected, the crankshaft must be checked for ovality as described in the following Section.
4   Failure of the big-end bearings is invariably accompanied by a pronounced knock within the crankcase. The knock will become progressively worse and vibration will also be experienced. It is essential that bearing failure is attended to without delay because if the engine is used on this condition there is a risk of more extensive damage occurring (ie breaking a connecting rod or even the crankshaft) due to the reduced oil pressure or through vibration.

5   Before the big-end bearings can be examined the bearing caps must be removed from each connecting rod. Each cap is retained by two high tensile bolts. Before removal, mark each cap in relation to its connecting rod so that it may be refitted correctly. As with the main bearings, wear will be evident in the form of scuffing or scoring and the bearing shells must be replaced as four complete sets.
6   Replacement bearing shells for either the big-end or main bearings are supplied on a selected fit basis (ie bearings are selected for correct tolerance to fit the original journal diameter), and it is essential that the parts to be used for renewal are of identical size.
7   Bearing shells must be selected in accordance with the size markings on both the connecting rod and crankshaft. See the following tables of sizes.
8   The relevant crankpin OD (outside diameter) code will be found on the edge of alternate flywheels. In the case of the crankpin, it is the **letter** (A or B) that is required, whilst the main bearing journal OD code appears as a **numeral** (1 or 2). The crankpin OD code is cross-referenced with the connecting rod code (1, 2 or 3) which is marked across the edge of the big-end eye. The main bearing OD code letter is cross-referenced to corresponding letters stamped at the rear of the crankcase (A, B or C) and will be found in the table below. Note that all crankshaft journals may also be checked by measuring with a micrometer. This method will permit the degree of wear and ovality to be assessed, by comparing the figures obtained with those indicated by the OD codes.

		CRANKPIN O. D. CODE	
		A	**B**
		34.996–35.006 mm	34.986–34.996 mm
CONNECTING ROD I. D. CODE	1	38.000–38.007 mm	D (Yellow)	C (Green)
	2	38.007–38.014 mm	C (Green)	B (Brown)
	3	38.014–38.021 mm	B (Brown)	A (Black)

Fig. 1.6 Big-end bearing shell selection table

		MAIN JOURNAL O. D. CODE	
		1	**2**
		32.990–33.000 mm	32.880–32.990 mm
CASE I. D. CODE	A 36.000–36.008 mm	D (Yellow)	C (Green)
	B 36.008–36.016 mm	C (Green)	B (Brown)
	C 36.016–36.024 mm	B (Brown)	A (Black)

Fig. 1.7 Main bearing shell selection table

9   The bearing shell thickness for both the main and big-end journals is colour-coded. The shells themselves are marked with with a dab of paint on one edge, the colours, and consequently the sizes, corresponding with those given in the tables above.

21.3 Check main bearing shells for damage and wear

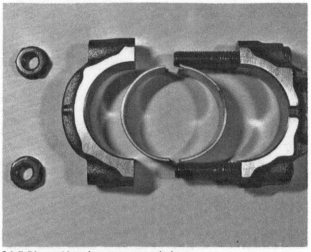
21.5 Big-end bearing set; general view

21.8a Crankshaft journal size letter on flywheel edge

21.8b Big-end journal size number on edge of big-end eye

22 Crankshaft assembly: examination and renovation

1   If the main or big-end bearing shells are found to be worn, the crankshaft journals should be checked with the aid of a micrometer, the journal material will not normally wear at anything like the rate of the soft bearing shells, but if the engine has been run for some time with worn bearings, ovality may develop. The manufacturer does not specify any ovality tolerance, but as a rough guide, ovality of 0.05 mm (0.002 in) or more will warrant remedial action.

2   The crankshaft should be checked for run-out by supporting it between lathe centres or on V-blocks. Arrange a dial gauge (DTI) to rest upon the centre main bearing journal, then rotate the crankshaft through two complete revolutions, noting the range shown on the gauge. This figure should then be halved to give the actual run-out figure. The service limit for crankshaft run-out is 0.05 mm (0.002 in).

3   The journal surface should be checked carefully for signs of scoring or damage, particularly when badly worn bearing shells have been discovered. If the crankshaft assembly is out of limits or damaged in any way it will be necessary to renew it. The manufacturer does not operate a service exchange scheme or supply undersize bearing shells, so reconditioning by re-grinding the bearing surfaces is not practicable. In some instances, independent engineering companies may be able to re-work a damaged crankshaft.

21.8c Connecting rod weight code marking letter

4   The clearance between any set of bearings and their respective journal may be checked by the use of Plastigauge (press gauge). Plastigauge is a graduated strip of plastic material that can be compressed between two mating surfaces. The resulting width of the material when measured with a micrometer will give the amount of clearance. For example if the clearance in the big-end bearing is to be measured, Plastigauge should be used in the following manner.

5   Cut a strip of Plastigauge to the width across the bearing to be measured. Place the Plastigauge strip across the bearing journal so that it is parallel with the crankshaft. Place the connecting rod complete with its half shell on the journal and then carefully replace the bearing cap complete with half shell onto the connecting rod bolts. Replace and tighten the retaining nuts to the correct torque and then loosen and remove the nuts and the bearing cap. Without bending or pressing the Plastigauge strip, place it as its thickest point between a micrometer and read off the measurement. This will indicate the precise clearance. The original size and wear limit of the crankshaft journals and the standard and service limit clearance between all the bearings is given in the specifications at the beginning of this Chapter.

6   The crankshaft has drilled oil passages which allow oil to be fed under pressure to the working surfaces. Care must be taken to clean these out carefully, preferably using compressed air.

7   When refitting the connecting rods and shell bearings, note that under no circumstances should the shells be adjusted with a shim, 'scraped in' or the fit 'corrected' by filing the connecting rod and bearing cap or by applying emery cloth to the bearing surface. Treatment such as this will end in disaster; if the bearing fit is not good, the parts connected have not been assembled correctly. This advice also applies to the main bearing shells. Use new big-end bolts too – the originals may have stretched and weakened.

8   Lubricate the bearing surfaces with molybdenum disulphide grease before reassembly takes place and make sure the tangs of the bearing shells are located correctly. The connecting rods must be fitted in their original locations and with the numerical code towards the front of the engine. After the initial tightening of the connecting rod nuts, check that each connecting rod revolves freely, then tighten to a torque setting of 2.4 – 2.8 kgf m 17 – 20 lbf ft. Check again that the bearing is quite free.

23 Connecting rods: examination and renovation

1   It is unlikely that any of the connecting rods will bend during normal usage, unless an unusual occurrence such as a dropped valve has caused the engine to lock. Carelessness when removing a tight gudgeon pin can also give rise to a similar problem. It is not advisable to straighten a bent connecting rod; renewal is the only satisfactory solution.

2   The small-end eye of the connecting rod is unbushed and it will be necessary to renew the connecting rod if the gudgeon pin becomes a slack fit. If the clearance between the gudgeon pin and the small-end is excessive, check first that the wear is in the eye and not the gudgeon pin. This will prevent the unnecessary renewal of a sound component. Always check that the oil hole in the small-end eye is not blocked since if the oil supply is cut off, the bearing surfaces will wear very rapidly.

24 Oil seals: examination and replacement

1   Oil seal failure is difficult to define precisely and as such it is difficult to give any firm recommendations. With the engine assembled and in the machine, the usual sign of a failed oil seal is in the form of oil showing on the outside of the engine.

2   The main oil seals (crankshafts and gearbox shafts) should always be renewed whenever the engine is dismantled. With oil seals being relatively inexpensive, it is advisable to renew all the seals as a matter of course when the engine is being overhauled. This will preclude any risk of an annoying oil leak developing after the unit has been reinstalled in the frame.

22.1 Check the crankshaft assembly for wear and damage

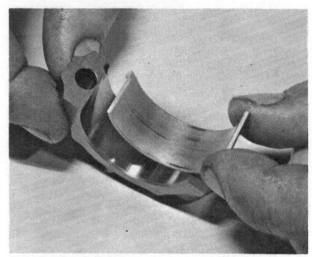

22.8a Bearing shell tangs must locate correctly

22.8b Torque the big-end bolts to the specified setting

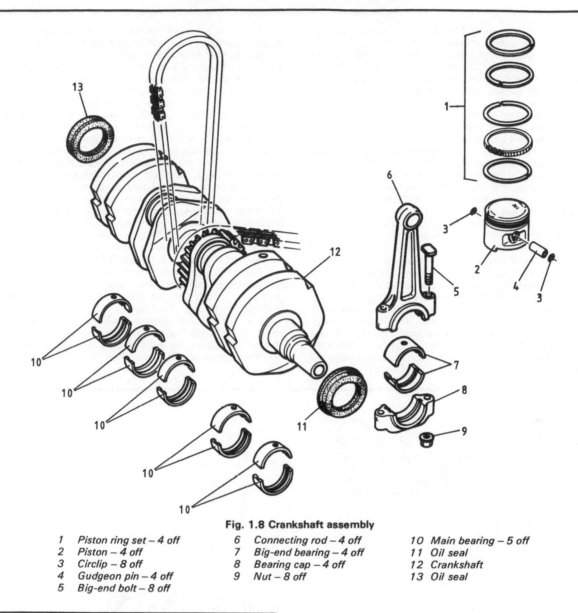

Fig. 1.8 Crankshaft assembly

1 Piston ring set – 4 off	6 Connecting rod – 4 off	10 Main bearing – 5 off
2 Piston – 4 off	7 Big-end bearing – 4 off	11 Oil seal
3 Circlip – 8 off	8 Bearing cap – 4 off	12 Crankshaft
4 Gudgeon pin – 4 off	9 Nut – 8 off	13 Oil seal
5 Big-end bolt – 8 off		

25 Primary shaft/starter clutch assembly, primary chain and tensioner: examination and renovation

1   Power from the crankshaft is transmitted to the primary shaft by way of a Morse, or Hy-Vo, chain. The primary shaft is supported at each end by a journal ball bearing, and incorporates a rubber-and-vane type shock absorber. Power from the primary shaft is taken off by a pinion on the right-hand end. This acts as a primary drive pinion, conducting power to the clutch. Also incorporated on the primary shaft, is the starter motor clutch. This device is fitted, together with the starter gear pinion, to the rear of the primary driven sprocket hub assembly. The starter gear runs on its own needle roller bearing.

2   The primary shaft journal ball bearings should be carefully washed out with petrol or a suitable cleaning solvent, then checked for wear or roughness by spinning them. Any rough spots or discernible radial play will indicate the need for renewal. The left-hand bearing will remain in the lower crankcase half when the shaft is removed. It can be drifted from position if renewal is required using a suitable drift, such as a large socket spanner. Take care during this procedure that the casing is not damaged through careless use of the hammer and drift. The right-hand (primary drive pinion side) bearing will have

remained on the shaft when it was removed from the casing as described in Section 17, paragraphs 5 and 6 of this Chapter. The bearing should not prove too tight a fit on the shaft and can often be removed by jarring the end of the shaft against a hardwood block.

3   Remove the needle roller bearing from the starter gear. The additional spacer may remain on the primary shaft or in the starter gear. Separate the starter gear from the back of the starter clutch. Examine the needle roller bearing carefully; replace the bearing if any damage is apparent.

4   The starter clutch device is retained on the back of the primary driven sprocket hub assembly by three bolts with curiously shaped heads. Honda term these bolts 'torx' bolts. In appearance they are an amalgamation of Allen bolts and crosshead screws. They require the use of a special impact driver bit, as recommended by Honda (Part No 07703-0010200), or alternatively, a modified Allen key, to release them. The modified Allen key may be cut and filed until it resembles the item shown in the accompanying photograph. The 'torx' bolts removal is further hampered by their being fitted tightly and secured with locking compound. If the home-made 'torx' bolt remover is to be used, it will be necessary to clamp the end of the key into the jaws of a self-locking wrench in order to obtain sufficient leverage to accomplish removal of the bolts.

5   Separate the starter clutch from the primary driven sprocket assembly with the three retaining bolts removed. Note the single locating dowel. Within the clutch are contained three rollers, three plungers and three springs. The only parts likely to require attention are the rollers and the springs. Signs of wear or damage will be obvious and will necessitate renewal of the worn or damaged parts.

6   Check the condition of the starter gear teeth. Also check the condition of the central spigot; the minimum permissible outside diameter is 41.93 mm (1.650 in). To check that the clutch is operating correctly, reassemble the bearing and spacer in the starter gear, and push the starter gear spigot back into position in the back of the clutch. Take care not to displace the bearing rollers. Turn the starter gear clockwise. This should force the spring loaded rollers against the clutch hub and cause it to tighten on the hub as the drive is taken up.

7   The primary shaft shock absorber consists of an outer drum with internal vanes which slides over similar vanes on the outside of a splined centre. Rubber damping blocks are fitted between each pair of vanes providing shock absorption in the event of snatch loadings in either direction. The assembly will not normally require attention until the rubber blocks become compressed after very high mileages have been covered.

8   If, however, the assembly does require attention it is possible to separate the assembly into its component parts. From the left-hand face of the primary driven sprocket, remove the large circlip, followed by the retaining collar and then the two large collet pieces. Remove the sprocket to expose the rubber damping blocks fitted to the splined centre.

9   Check the rubbers for damage or compression. If the unit is in good condition, there will be no discernible free play when it is assembled, and the rubbers should make the two parts a tight fit. Any slackness will allow snatch loadings to be transferred to the clutch and gearbox, making the machine rather unpleasant to ride. The unit is reassembled in the reverse order of dismantling. If new rubbers prove to be a very tight fit, use a small amount of petrol as a lubricant. This will make assembly much easier, the petrol evaporating off soon afterwards.

10   The Morse type primary chain is automatically tensioned by a spring-loaded tensioner mechanism. Examine the tensioner slipper blade for damage or excessive wear. If the Teflon coating on the blade is worn through, the blade should be renewed. This type of chain is very resistant to stretching, and very high mileages can normally be covered before renewal is necessary. The chain should be checked for wear whenever the engine is stripped for overhaul, following the procedure outlined below. If at or near the service limit, it is worthwhile renewing the chain in view of the considerable amount of dismantling work that will be required should it prove worn out in the near future.

11   Assemble the chain around the crankshaft and primary shaft sprockets, anchoring the crankshaft against suitable stops on the workbench. Attach a spring balance to the primary shaft and apply a tension of 32 kg (71 lb) using a spring balance. With the chain under tension, measure the chain length as shown in Fig. 1.10. The nominal length is 114.15 – 114.40 mm (4.494 – 4.504 in). The chain must be renewed when it reaches the service limit of 115.5 mm (4.55 in).

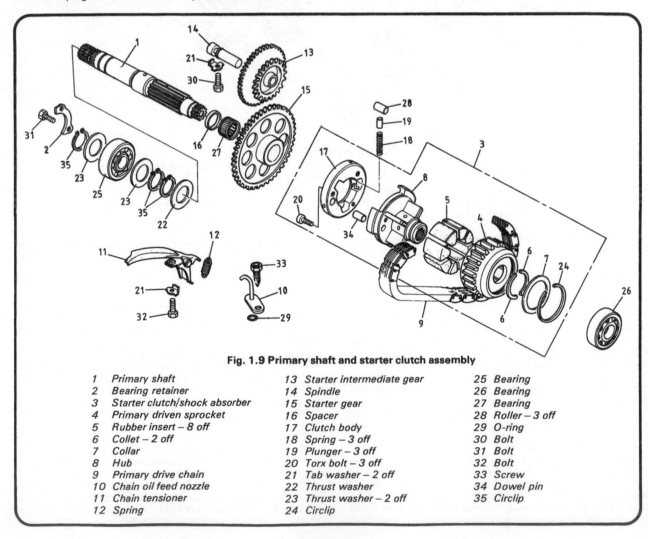

Fig. 1.9 Primary shaft and starter clutch assembly

1 Primary shaft	13 Starter intermediate gear	25 Bearing
2 Bearing retainer	14 Spindle	26 Bearing
3 Starter clutch/shock absorber	15 Starter gear	27 Bearing
4 Primary driven sprocket	16 Spacer	28 Roller – 3 off
5 Rubber insert – 8 off	17 Clutch body	29 O-ring
6 Collet – 2 off	18 Spring – 3 off	30 Bolt
7 Collar	19 Plunger – 3 off	31 Bolt
8 Hub	20 Torx bolt – 3 off	32 Bolt
9 Primary drive chain	21 Tab washer – 2 off	33 Screw
10 Chain oil feed nozzle	22 Thrust washer	34 Dowel pin
11 Chain tensioner	23 Thrust washer – 2 off	35 Circlip
12 Spring	24 Circlip	

25.2a Primary shaft left-hand bearing will remain in the lower crankcase after shaft removal

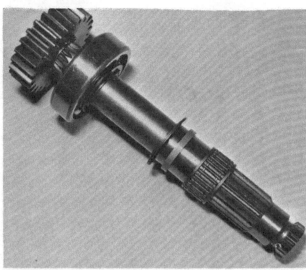

25.2b Primary shaft: general view (pinion shown in position for reference purposes)

25.4a Special tool may be fabricated to release 'torx' bolts

25.4b Using the special tool, remove the 'torx' bolts and ...

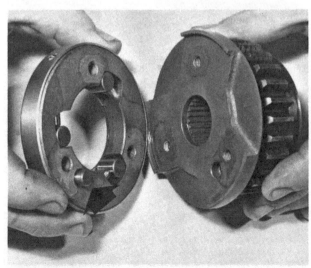

25.5a ...separate the starter clutch from the sprocket assembly. Note the locating dowel (arrowed)

25.5b Remove the rollers only if examination is required

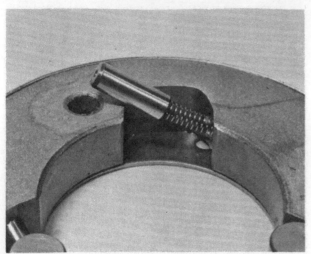

25.5c Plunger and spring will be released when roller is removed. Check spring for wear

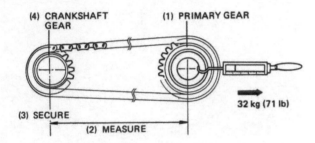

(4) CRANKSHAFT GEAR

(1) PRIMARY GEAR

(3) SECURE

(2) MEASURE

32 kg (71 lb)

Fig. 1.10 Primary chain wear measurement

26 Cylinder block: examination and renovation

1   The usual indication of badly worn cylinder bores and pistons is excessive smoking from the exhausts. This usually takes the form of blue haze tending to develop into a white haze as the wear becomes more pronounced.

2   The other indication is piston slap, a form of metallic rattle which occurs when there is little load on the engine. If the top of the bore is examined carefully, it will be found that there is a ridge on the thrust side, the depth of which will vary according to the rate of wear which has taken place. This marks the limit of travel of the top piston ring.

3   Measure the bore diameter just below the ridge using an internal micrometer, or a dial gauge. Compare the reading you obtain with the reading at the bottom of the cylinder bore, which has not been subjected to any piston wear. If the difference in readings exceeds 0.10 mm (0.004 in) or the maximum bore diameter exceeds 59.90 mm (2.358 in) the cylinder block will require boring and honing to the next oversize.

4   If measuring instruments are not available, the amount of bore wear can be approximated as follows. Remove the rings from one position (see the following Section), then slide it into

its bore so that the crown is about $\frac{3}{4}$ in from the top. Using feeler gauges, measure the gap between the piston and the bore at 90° to the gudgeon pin boss. If the gap exceeds 0.10 mm (0.004 in) remedial action may be required.

5   If wear has necessitated re-boring, the work should be entrusted to a Honda Service Agent or to a competent engineering shop.

6   Honda can supply pistons in two oversizes only: + 0.25 mm and + 0.50 mm.

7   Check that the surface of the cylinder bores is free from score marks or other damage that may have resulted from an earlier engine seizure or a displaced gudgeon pin. A rebore will be necessary to remove any deep scores, irrespective of the amount of bore wear that has taken place, otherwise a compression leak will occur.

8   Make sure the external cooling fins of the cylinder block are free from oil and road dirt, as this can prevent the free flow of air over the engine and cause overheating problems.

27 Pistons and piston rings: examination and renovation

1   If a rebore becomes necessary, the existing pistons and piston rings can be discarded becuase they will have to be replaced by their new oversizes.

2   If a rebore is not considered necessary, examine each piston closely. Reject pistons that are scored or badly discoloured as the result of exhaust gases by-passing the rings.

3   Remove all traces of carbon from the piston crowns, using a blunt ended scraper to avoid scratching the surface. Finish off by polishing the crowns of each piston with metal polish, so that carbon will not adhere so rapidly in the future. Never use emery cloth on the soft aluminium.

4   Piston wear usually occurs at the skirt or lower end of the piston and takes the form of vertical streaks or score marks on the thrust side of the piston. Damage of this nature will necessitate renewal, if severe.

5   The piston ring grooves may become enlarged in use, allowing the rings to have a greater side float. If the clearance exceeds 0.15 mm (0.006 in) the pistons are due for replacement.

6   To measure the end gap, insert each piston ring into its cylinder bore, using the crown of the bare piston to locate it about 1 inch from the top of the bore. Make sure it is square in the bore and insert a feeler gauge in the end gap of the ring. If the gap is outside the wear limit, the ring(s) concerned must be renewed.

Piston ring end gap
  Nominal ring end gap

Top and 2nd	0.10-0.30 mm (0.004-0.012 in)
Oil control	0.30-0.90 mm (0.012-0.035 in)
Wear limit	
Top and 2nd	0.7 mm (0.028 in)
Oil control	1.1 mm (0.043 in)

7   Check that there is no build up of carbon on the inside surface of the rings or in the grooves of the pistons. Any build up should be removed by careful scraping. An old, broken ring is useful for this.

8   When refitting new piston rings, it is also necessary to check the end gap. If there is insufficient clearance, the rings will break up in the bore whilst the engine is running and cause extensive damage. The ring gap may be increased by filing the ends of the rings with a fine file, though this is not normally necessary with new rings of Honda manufacture.

9   The ring should be supported on the end as much as possible to avoid breakage when filing, and should be filed square with the end. Remove only a small amount of metal at a time and keep rechecking the clearance in the bore.

10   When dealing with piston rings it is advisable to attend to one piston at a time to preclude the risk of rings being refitted to the wrong piston. This would be undesirable as the rings will have bedded into a particular bore, and would allow com-

pression leakage if fitted incorrectly. It will be noted that the two compression rings differ in that the top ring has a plain profile, whilst the second ring is tapered. Each ring is marked on one face and this should be arranged to face upwards when fitted. The oil control ring is built up from two scraper rails separated by an expander. Rings from one manufacturer only should be used on any one piston.

11 When installing the piston rings, or when removing sound existing rings for examination, care should be taken not to break them by spreading the ends too far apart. With a little experience the ring gap can be eased apart by hand, just enough to permit removal or fitting. Alternatively, three or four thin metal strips can be placed across the ring grooves and the rings slid on or off (See the accompanying illustration).

12 The piston crowns will show whether the engine has been rebored on some previous occasion. All oversize pistons have the rebore size stamped on the crown. This information is essential when ordering replacement piston rings.

13 A replacement set of rings is relatively inexpensive and it is considered good practice to renew them as a matter of course during a major overhaul.

27.3 Clean piston crowns carefully and thoroughly; note 'IN' mark provided to aid correct positioning

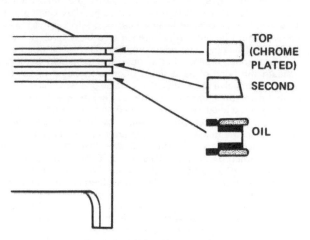

Fig. 1.11 Piston ring profiles

27.5 Using a feeler gauge, check the wear on the piston ring grooves and ..

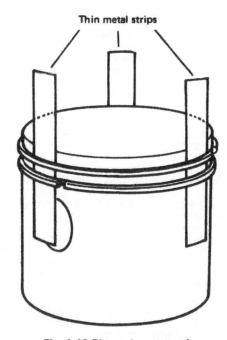

Fig. 1.12 Piston ring removal

27.6 ... the piston ring end gap

28 Cylinder head: examination and renovation

1   Remove all traces of carbon from the cylinder head using a blunt ended scraper (the round end of an old steel rule will do). Finish by polishing with metal polish to give a smooth shiny surface. This will aid gas flow and will also prevent carbon from adhering so firmly in the future.
2   Check the condition of the sparking plug hole threads. If the threads are worn or crossed they can be reclaimed by a Helicoil insert. Most motorcycle dealers operate this service which is very simply, cheap and effective.
3   Clean the cylinder head fins with a wire brush, to prevent overheating, through dirt blocking the fins.
4   Lay the cylinder head on a sheet of $\frac{1}{4}$ inch plate glass to check for distortion. Aluminium alloy cylinder heads distort very easily, especially if the cylinder head bolts are tightened down unevenly. If the amount of distortion is only slight, it is permissible to rub the head down until it is flat once again by wrapping a sheet of very fine emery paper (400-600 grade) around the plate glass base and rubbing with a rotary motion.
5   If it proves possible to insert a 0.25 mm (0.010 in) feeler gauge between the glass plate and the cylinder head, the head is beyond the service limit for distortion and remedial action must be taken. If only just outside this limit it may be possible to have the gasket face machined flat, but great care must be taken if this action is chosen. Remember that the machining operation will have some effect on the compression ratio, and might result in the valves touching the piston at high speed. A specialist machinist should be sought for this type of work.

29 Valves, valve seats, and valve guides: dismantling, examination and renovation

1   It is best to remove all carbon deposits from the combustion chambers, as described in paragraph 1 of the previous Section, before removing the valves.
2   Before the valves can be removed, it is necessary to obtain a valve spring compressor of the correct size. Compress each set of valve springs in turn, and remove the split collets from the valve cap, and then the valve and valve spring assembly. Also remove the inner and outer springs lower seats, and the oil seal from the valve guides. The oil seals on the valve guides should be renewed upon reassembly. Do not compress the springs more than is necessary to facilitate removal of the split collet halves.
3   As each valve and its associated parts are released, place them in a suitably marked bag or container to ensure that they are refitted in their correct location. If this precaution is not observed, compression leakage will be almost inevitable. When cleaning and examining these components deal with one valve assembly at a time, for the same reason.
4   Before giving the valves and valve seats further attention, check the clearance between each valve stem and the guide in which it operates.
5   The amount of valve stem wear must be checked by measuring the stem with a micrometer. Check the stem in a number of positions, and note the smallest reading obtained. If this falls below the service limit, the valve must be renewed.

Inlet	Exhaust
5.47 mm (0.215 in)	5.45 mm (0.214 in)

6   The internal diameter of the valve guides can be measured by using ball gauges or a suitable inside micrometer. The service limits for the guides are as follows.
Valve guide ID service limit

Inlet	Exhaust
5.55 mm (0.219 in)	5.55 mm (0.219 in)

7   The valve stem diameter figure should now be subtracted from the valve guide ID figure to give the amount of valve to guide clearance. This should not exceed 0.08 mm (0.003 in) for the inlet valve or 0.10 mm (0.004 in) for the exhaust valve. If the total clearance exceeds this, check whether it can be brought within tolerance by fitting a new valve only. Failing this, a new guide and valve must be fitted. In either case, remember that the valve seat must be re-cut.
8   In the correct measuring equipment is not available, check for wear by placing the valve in its guide and attempting to rock the valve to and fro. Anything more than barely discernible play will be indicative of unacceptable wear. If present, take the cylinder head and valves to a Honda Service Agent for verification of wear.
9   The valve guides are an interference fit in the cylinder head, and will require a stepped drift and a certain amount of skill during the renewal operation. The stepped drift should have a spigot of similar size to the valve stem, whilst the larger diameter should be slightly less than that of the valve guide. Before attempting to remove the old valve guide, clean any carbon deposits from the port side of the valves. This will ease removal and help prevent broaching of the valve guide housing bores. To facilitate guide removal, place the cylinder head in an oven and heat it to about150°C. Remove the head from the oven, place it on blocks, so that the combustion chambers face upwards, and drive the old guide(s) out of the head casting. Note that to prevent distortion of the large alloy casting, it is essential that the cylinder head is heated evenly. For this reason an oven **must** be used in preference to a blow torch or other methods of heating. The new component(s) can be fitted in a similar manner, noting that a new O-ring should be fitted beneath the head of each guide prior to its being fitted. The replacement valve guide should always be one which is oversize so that it can be reamed out to the correct size after it has been inserted in the head. Great care should be taken during both the removal and replacement operations, as the soft alloy head casting and the guides themselves are easily damaged.
10 The guides must be reamed to size after fitting, using the Honda reamer, part number 07984 – 2000000, or equivalent, to bring the ID to 5.55 mm (0.219 in). The valve seats must now be re-cut to the new guide as described in paragraphs 16, 17 and 18.
11 Each valve should be cleaned prior to checking for wear. Carbon deposits should be removed from the top and underside of the head, taking care not to score the seating face or stem. Remove the heavy carbon deposits using a blunt ended scraper and then finish off with a fine emery paper. When cleaning the stem of the valve use only longitudinal strokes and not rotating strokes. The fine scoring caused by the emery paper may cause stress failure if it runs around the circumference of the valve stem. A highly polished finish is desirable because it reduces the rate of carbon build-up in the future.
12 Examine the valve seating face in conjunction with the valve seat in the cylinder head, looking for signs of pitting or burning. This is most likely to be found on the exhaust valves, because these lead an altogether more strenuous life than their inlet counterparts. If the machine has been maintained properly, there should be no more than a few minor marks on either face, but if severe damage is discovered, remedial action will be required. Small marks can be removed by lapping as described below.
13 Valve grinding-in will be necessary, irrespective of whether new guides have been fitted, to remove the marks described above.
14 Valve grinding is a simple task. Commence by smearing a trace of fine valve compound (carborundum paste) on the valve seat and apply a suction tool to the head of the valve. Oil the valve stem and insert the valve in the guide so that the two surfaces to be ground in make contact with one another. With a semi-rotary motion, grind in the valve head to the seat, using a backward and forward action. Lift the valve occasionally so that the grinding compound is distributed evenly. Repeat the application until an unbroken ring of light grey matt finish is obtained on both valve and seat. This denotes the grinding operation is now complete. Before passing to the next valve, make sure that all traces of the valve grinding compound have been removed from both the valve and its seat and that none

has entered the valve guide. If this precaution is not observed, rapid wear will take place due to the highly abrasive nature of the carborundum paste.

15 If a reasonable amount of lapping fails to produce the required unbroken seating area on both the valve and seat, the operation must be abandoned. Excessive lapping is time-consuming and will only result in a ruined valve seat. Examine the seating faces very closely. The attempt at lapping will have highlighted any pits, and a decision must now be taken on the best course of action. Honda advise that the valve should **not** be re-faced, and must therefore be renewed if damaged to the extent that lapping proves ineffective.

16 The valve seats may be re-cut to remove pitting or to compensate for poor seating or incorrect seat widths. Note that if new valve guides have been fitted, the valve seats will need re-cutting to realign the seats with the guide axis. The valve seats are formed by cutting them at three angles to produce the correct seat width, which is nominally 1.2 mm (0.047 in). It should be noted that there is nothing to be gained by using an excessively large contact area, as this will lead to accelerated wear and pitting, and will impair gas flow through the engine.

17 The re-cutting operation requires the use of three separate cutters of various angles and diameters. These are shown in the accompanying diagram. The 32° and 60° cutters are used first, and the correct seating width is then obtained by using the 45° cutter. A word of caution is necessary here, because the valve seats will only accept a limited amount of cutting before they become unacceptably pocketed. When this happens, the seat is no longer usable, and as it is integral with the cylinder head this will require renewal. In view of this risk, and the cost of cutters (not to mention new cylinder heads) it is strongly recommended that any re-cutting is entrusted to a Honda Service Agent. The above information is therefore given for the benefit of owners having access to the required tools and the skill to use them safely.

18 The newly-cut seats should be carefully lapped to their respective valves as described in paragraphs 14 and 15 above.

All being well, a perfect gas-tight joint should result. If not, it is likely that the seating face of the valve is unusable, and the valve must be renewed.

19 In view of the amount of skilled work involved in cylinder head reconditioning, some thought should be given to the alternative of entrusting the job to a Honda Service Agent. Bear in mind that some specialist equipment is needed, and is unlikely to warrant purchasing for a one-off job. Much of the cost involved is in the stripping and reassembly work which is a necessary precursor to overhaul. If this stage is undertaken at home, a good proportion of the total cost can be saved.

20 Before reassembling the valves, check the spring seats, springs and collet halves for signs of wear or damage. The valve springs will take a permanent set after very high mileages, and will eventually allow valve float to occur at high speeds. The free lengths of the springs should be measured with a vernier caliper and compared with the table below. Springs are relatively cheap when compared with the cost of rectifying the damage that would result from a valve head hitting a piston at high engine speeds – if in doubt, play safe and renew them.

Valve spring minimum free length
Inner	37.9 mm (1.49 in)
Outer	43.3 mm (1.70 in)

21 Reassemble the valves and valve springs by reversing the dismantling procedure. Ensure both the inner and outer springs lower seats are refitted. Fit new oil seals to each valve guide and oil both the valve stem and the valve guide, prior to reassembly. Take special care to ensure the valve guide oil seal is not damaged when the valve is inserted. On inspection it will be seen that the valve spring coils are more closely wound at one end of the spring than the other. When installing the springs ensure that the close wound end faces the cylinder head. As a final check after assembly, give the end of each valve stem a sharp tap with a hammer, to make sure the split collets have located correctly.

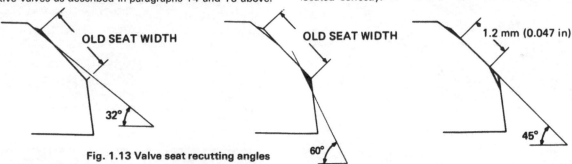

Fig. 1.13 Valve seat recutting angles

29.2 Compress the valve springs and remove the split collets

29.21a Fit new oil seals to each valve guide

29.21b Lubricate the valve stem thoroughly before insertion

29.21c Ensure that the larger, outer spring, seat and then ...

29.21d ... the smaller, inner spring, seat are refitted

29.21e Refit both springs to each valve; closer wound coils downwards

29.21f Refit the valve collar

29.21g Recompress the springs and fit the collets to secure the assembly

30 Rocker spindles and rocker arms: examination and renovation

1 Remove the four rocker spindles from the cylinder head cover (rocker cover). With the cover positioned so that the outer face is uppermost, remove the four nuts from the four cotter pins which secure the inner ends of each spindle. Drive out the cotter pins. Remove the Allen bolts from the outer end of each rocker spindle. Note the green O-ring fitted to each bolt. The spindle may now be withdrawn from the cover. To facilitate spindle removal, insert a suitable 6 mm bolt (an engine casing bolt will do admirably) into the threaded outer end of the spindle and pull it out. As the spindles are withdrawn, the rocker arms and their springs will be freed allowing their removal. As each rocker arm and its spring is released and removed, place it in a suitably marked bag or container to ensure that it is refitted in its correct location upon reassembly.

2 Check the rocker arms for excessive clearance when fitted on the rocker spindle. This is indicated by slackness between the rocker arm and spindle. The service limit for the internal diameter of the rocker arm bore is 12.05 mm (0.474 in). Each rocker spindle should also be examined and measured. The

maximum permissible wear has occurred when the spindle outside diameter is 11.94 mm (0.470 in). If one or both of these components has reached the service limit given, they must be replaced with new items.

3 Examine the end of each rocker arm where it bears directly on the operating cam and renew any showing signs of scuffing, cracking or breakthrough into the specially hardened surface. At the other end of the rocker arm, the internal thread into which the valve adjuster screws must be in good condition, also that of the adjuster itself and the locknut. The tip of the adjuster is hardened; if the tip is cracked, broken or badly scuffed, the complete adjuster must be renewed.

4 Reassembly of the rocker components should be made by reversing the dismantling procedure. Ensure that the rocker spindles and rocker arms are replaced in their original positions in the cylinder head cover. The spindles must be fitted with the threaded end outermost. The cotter pin retaining nuts should be tightened down to a torque setting of 1.0 – 1.4 kgf m (7 – 10 lbf ft). Fit new O-rings to each of the four Allen bolts which secure the outer ends of the rocker spindles. These Allen bolts should be torqued down to a setting of 0.8 – 1.2 kgf m (6.9 lbf ft). Oil all components liberally during reassembly; the rocker arms must move freely once reassembled.

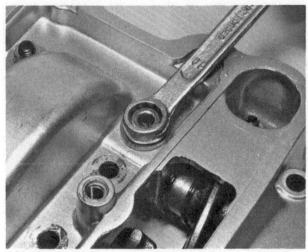

30.1a Remove the four nuts from the top of the rocker cover and ...

30.1b ... remove the four cotter pins

30.1c Remove the Allen screws (note the green O-rings)

30.1d Insert a 6mm bolt in the spindle end and ...

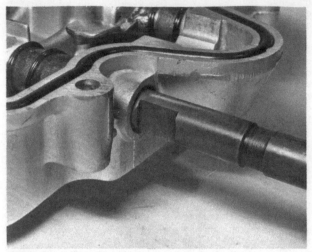

30.1e ... withdraw the spindle

30.1f The rocker arms and springs are now free to be removed

31 Camshaft, camshaft drive chain and tensioner: examination and renovation

1   The camshaft is carried directly in the cylinder head and is retained in position by camshaft bearing surfaces in the cylinder head cover.

2   Examine the camshaft bearing surfaces for signs of wear or scoring. This type of damage should not occur, because the bearing surfaces are supplied with copious amounts of engine oil during normal running. If, however, the oil has not been changed regularly and the filter has been allowed to become blocked, the bypass system will circulate unfiltered oil carrying abrasive dirt particles to the bearings. The damage that this may cause will be most evident on the soft alloy bearing surfaces in the cylinder head and cylinder head cover. Little can be done in these cases unless a specialist engineering firm, of the type advertised in the motorcycle press, can be found willing and capable of refurbishing these bearing surfaces. If such a firm cannot be found the only alternative course of action is to renew the cylinder head and cover, and the camshaft if this too is scored. It is worthwhile at this juncture to emphasise the importance of oil changes at the recommended intervals. It is clearly false economy to try and extend the period between oil changes, the recommended interval of 4000 miles (6500 km) must be adhered to.

3   Camshaft runout can be checked by supporting each end of the camshaft on V-blocks and arranging a dial gauge to bear upon the centre bearing journal. Runout must not exceed 0.1 mm (0.004 in). Remember that the out of true figure is actually half that of the runout reading observed.

4   When the camshaft is in position in the cylinder head and the head cover is installed and tightened down to its correct torque setting, there should be a specific oil clearance between the camshaft and the cylinder head cover bearing surfaces. This clearance can be measured by using plastigauge in the same way as described in Section 22 of this Chapter. The nominal camshaft to head cover bearing surface clearance is 0.160 – 0.202 mm (0.0063 – 0.0080 in) with a service limit of 0.21 mm (0.008 in).

5   The camshaft and head cover bearing surfaces should be free from oil prior to measurement. Place a strip of plastigauge on top of each camshaft journal, fit the head cover and retaining bolts, and tighten down to the recommended torque setting of

0.8 – 1.2 kgf m (6 – 9 lbf ft), in a diagonal pattern to preclude warpage. If the clearance is outside the service limit, check whether renewal of the camshaft will suffice to bring it within limits. Failing this, either a reconditioned head and head cover must be obtained or the cylinder head and head cover must be renewed.

6   Examine each of the camshaft lobes for signs of wear or scoring. A worn camshaft will show signs of flats developing on the peak of each lobe, and the degree of wear can be checked by measuring the lobe across its highest point. The service limit for the camshaft inlet valve lobes is 35.6 mm (1.40 in), whilst that of the camshaft exhaust valve lobes is 35.3 mm (1.39 in). Also examine the camshaft closely for evidence of cracking, scuffing or surface breakthrough of any of the cam lobes; damage of this kind will necessitate renewal.

7   The camshaft is driven by a narrow Hy-Vo type chain.

8   The Hy-Vo type chains are well known for their resistance to stretching, but eventually some wear will take place with the ultimate result that the tensioner will be unable to compensate for the increased length. The chain length should therefore be checked whenever it is removed. If the chain is at or near the service limit it is advisable to renew it to avoid further dismantling work in the near future.

9   Arrange the chain around the camshaft sprocket and the crankshaft, anchoring one of these to the workbench. Using a spring balance apply a pressure of 13 kg (29 lb) and measure the distance shown in Fig. 1.15. If this exceeds 186.4 mm (7.34 in) the chain has passed the service limit and must be renewed.

10   The tensioner mechanism consists of a spring-loaded blade which takes up any normal chain slack. Tension is set automatically by releasing the tensioner lock nut with the engine not operating, to allow the tensioner to assume its correct position, the lock nut then being used to secure the setting. On a high mileage engine, wear in the cam chain and tensioner will allow a lot of mechanical noise to develop. If the engine was noisy prior to dismantling, despite attempts at tensioner adjustment, check the tensioner's rubbing surface for wear. The tensioner 'slipper' blade is coated with a Teflon coating. If this coating is worn away, or the blade is deeply scored or damaged in any other way, it should be renewed in conjunction with the chain guide blade. Examine the condition of the tensioner spring; renew if suspect. Note that a new tensioner assembly will not compensate for a worn-out chain.

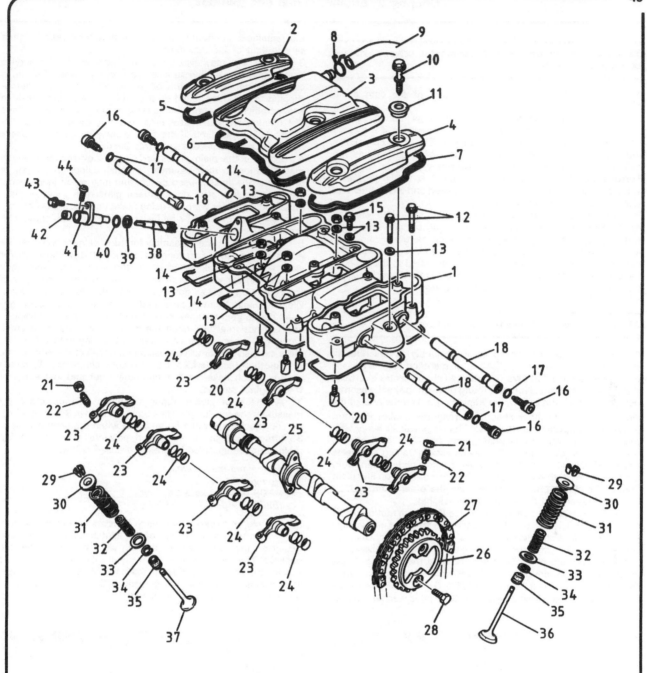

Fig. 1.14 Cylinder head cover and valve gear

1 Cylinder head cover	16 Allen screw – 4 off	31 Outer spring – 8 off
2 Right-hand inspection cover	17 O-ring – 4 off	32 Inner spring – 8 off
3 Breather cover	18 Rocker spindle – 4 off	33 Outer spring seat – 8 off
4 Left-hand inspection cover	19 Seal	34 Inner spring seat – 8 off
5 Seal	20 Cotter pin – 4 off	35 Oil seal – 8 off
6 Seal	21 Lock nut – 8 off	36 Inlet valve – 4 off
7 Seal	22 Adjuster screw – 8 off	37 Exhaust valve – 4 off
8 Clip	23 Rocker – 8 off	38 Tachometer drive shaft
9 Breather hose	24 Spring – 8 off	39 Thrust washer
10 Bolt – 6 off	25 Camshaft	40 O-ring
11 Seal – 6 off	26 Sprocket	41 Housing
12 Bolt – 6 off	27 Cam chain	42 Oil seal
13 Washer – 18 off	28 Bolt – 2 off	43 Bolt – 2 off
14 Nut – 4 off	29 Split collet – 16 off	44 Screw
15 Bolt – 8 off	30 Collar – 8 off	

32 Cam chain sprockets and tachometer drive: examination and renovation

1   Do not omit to check the condition of the cam chain sprockets.

2   The camshaft upper chain sprocket is bolted to the camshaft and in consequence is easily renewable if the teeth become hooked, worn, chipped or broken. The lower sprocket is integral with the crankshaft and if any of these defects are evident, the complete crankshaft assembly must be renewed. Fortunately, this drastic course of action is rarely necessary since the parts concerned are fully enclosed and well lubricated, working under ideal conditions.

3   If the sprockets are renewed, the chain should be renewed at the same time. It is bad practice to run old and new parts together since the rate of wear will be accelerated.

4   The worm drive to the tachometer is an integral part of the camshaft which meshes with a pinion attached to the cylinder head cover. If the worm is damaged or badly worn, it will be necessary to renew the camshaft complete.

33 Clutch assembly: examination and renovation

1   With the clutch removed as an assembly as described in Section 12 of this Chapter, further dismantling can take place on the workbench. Lift away the flanged clutch centre to expose the first friction plate. Note the fitting of a spring shock absorber device between the outermost friction plate and the rear of the clutch centre. This shock absorber arrangement takes the form of two rings of unequal thickness. The rings will lift away with the clutch centre. The heavier gauge ring is a seat for the thinner ring to bear upon. The device assists freeing and helps absorb shocks. This first, outermost, friction plate differs from the remainder in that it is thicker. Lift away the friction and plain plates, the clutch pressure plate, then the outer drum with its internal spacer, and finally the thrust washer.

2   After an extended period of service the clutch friction plates will wear sufficiently to allow clutch slip to occur under high loads. The friction plates should be checked for wear by measuring across the lining material using a vernier caliper. The nominal thickness of all but the outermost friction plate (plate B) is 2.62 – 2.78 mm (0.103 – 0.109 in). The nominal thickness for plate B is 3.42 – 3.58 mm (0.135 – 0.141 in). Renew plate B when the service limit of 3.2 mm (0.13 in) is reached. Renew the remainder of the friction plates if they have reached their service limit of 2.4 mm (0.09 in).

3   Clutch slip may also occur on plates within the service limits where the friction surface has become glazed. Glazing is caused by deposits of burnt oil generated by frequent clutch-slipping and is often found on machines ridden consistently in heavy traffic. It is permissible to remove the glazing by judicious use of abrasive paper assuming that this does not take the plates below their limit. If this fails to solve the problem, check the springs as detailed later in this Section.

4   Examine the plain plates for signs of overheating, which may have led to their assuming a bluish colour. This again is indicative of a heavily-used clutch, and may result in warpage. Place each plain plate on a surface plate or a sheet of plate glass and check for warpage using feeler gauges. If it is possible to insert a 0.3 mm (0.012 in) feeler gauge, the plate(s) must be renewed.

5   Examine the clutch assembly for burrs or indentation on the edges of the protruding tongues of the inserted plates and/or slots worn in the edges of the outer drum with which they engage. Similar wear can occur between the inner tongues of the plain clutch plates and the slots in the clutch inner drum. Wear of this nature will cause clutch drag and slow disengagement during gear changes, since the plates will become trapped and will not free fully when the clutch is withdrawn. A small amount of wear can be corrected by dressing with a fine file; more extensive wear will necessitate renewal of the worn parts.

6   Another factor which can promote clutch slip is clutch springs which have taken a permanent set and lost some of their original tension.

7   A check of clutch spring condition can be made by measuring the free length. The nominal length is 36.8 mm (1.45 in) and the springs must be renewed when the service limit of 35.4 mm (1.39 in) is reached. If possible, check the spring preload at the specified length, as detailed below.

Spring preload measurement
Nominal
27.08 – 29.93 kg @ 24.1 mm
(59.70 – 65.98 lb @ 0.95 in)

8   The clutch release mechanism consists of a cam-operated lever mounted on the inside of the clutch outer casing, and acting on the clutch release bearing via a mushroom-headed pushrod. The mechanism is relatively sturdy and is unlikely to require attention other than in the case of obvious breakage. The moving parts can be greased whenever the cover is removed for attention to the clutch.

32.4 Tachometer drive pinion in the cylinder head cover

33.1a Lift away the clutch centre and ...

33.1b ... the spring shock absorber device fitted to the rear

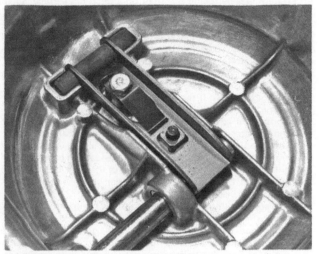

33.8 Grease all moving parts of clutch release mechanism in the outer cover

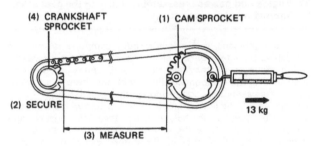

Fig. 1.15 Cam chain wear measurement

(4) CRANKSHAFT SPROCKET
(1) CAM SPROCKET
(2) SECURE
(3) MEASURE
13 kg

34 Gearbox components: examination and renovation

1   Examine each of the gear pinions to ensure that there are no chipped or broken teeth and that the dogs on the end of the pinions are not rounded. Gear pinions with any of these defects must be renewed; there is no satisfactory method of reclaiming them. General wear will lead to backlash developing between the pairs of gears, and this is difficult to check visually. If suspected, check the backlash as described in Section 19.

2   Dismantle each of the gearbox shafts in turn, laying out the components in the order in which they were removed. In the case of the mainshaft (input shaft), the 1st gear pinion is integral, but the remaining gears can be removed after releasing the circlips and washers which secure them. The layshaft (output shaft) can be dismantled in a similar manner, leaving the ball journal bearing in position.

3   Check the internal diameter of each gear pinion, comparing the readings obtained with the service limits listed below. An internal micrometer or vernier caliper will be required for this operation.

Gearbox pinion internal diameters – service limits
    Mainshaft 4th gear 25.06 mm (0.987 in)
    Mainshaft 5th gear 23.06 mm (0.908 in)
    Layshaft 1st gear 24.06 mm (0.947 in)
    Layshaft 2nd and 3rd gear 25.07 mm (0.987 in)

4   Check the various gear bushes for wear by measuring the internal diameter (ID) and outside diameter (OD). Renew if beyond the service limit shown below.

Gear bush ID and OD – service limits
    Mainshaft 5th gear (OD) 22.96 mm (0.904 in)
    Mainshaft 5th gear (ID) 20.07 mm (0.790 in)
    Layshaft 1st gear (OD) 23.95 mm (0.943 in)
    Layshaft 1st gear (ID) 20.07 mm (0.790 in)

5   Measure the outside diameters of the mainshaft and layshaft in the positions indicated in Fig. 1.16. Compare the readings obtained with the service limits shown below.

Mainshaft and layshaft OD – service limits
    A 24.93 mm (0.981 in)
    B and C 19.93 mm (0.785 in)

6   The clearance between each gearbox pinion and the relevant shaft or bush can be calculated from the above, in each case the clearance wear limit is 0.10 mm (0.004 in).

7   The gearbox bearings should be washed carefully in clean petrol and allowed to dry. The condition of the bearings can be checked by spinning them to highlight any roughness. Any discernible radial play is indicative of the need for renewal. Very slight axial play is normal and thus acceptable, but if excessive will require renewal. Inspect the bearing tracks and the balls or rollers. These will be unmarked and highly polished in a sound bearing, any surface defects indicate that the bearing is worn out. Retain any sound bearing, and lubricate it with oil to prevent corrosion forming prior to reassembly.

35 Gear selector mechanism: examination and renovation

1   Examine the selector forks for wear, noting the surface finish around the fork ends where they engage with the gearbox pinions. Wear here is unlikely unless lubrication has been badly neglected, in which case scoring may be evident. The width of the fork should be measured for wear, renewal being necessary if the reading obtained is less than 5.6 mm (0.22 in). Check the internal bore measurement of each fork. This should not be greater than 38.075 mm (1.499 in) for the larger gearchange drum mounted fork, and 13.04 mm (0.513 in) for the two smaller forks fitted to the support shaft.

2   The selector fork support shaft should be examined for signs of wear or scoring, and rejected if badly damaged. Check

for straightness by rolling the shaft on a dead flat surface, such as a surface plate or a sheet of glass. If at all bent, the shaft should be renewed as it will seriously impair gear selection if refitted. The outside diameter of the shaft can be measured for wear if there is some doubt as to its condition, it must be renewed if worn to 12.90 mm (0.508 in) at any point. Check the fit of the shaft in its casing bore. It should normally be a light sliding fit. Excessive wear will allow movement and thus sloppy gearchanging action. If the casing has become worn, it may be necessary to have it bored out and bushed. This should be left to a competent engineering company.

3   The selector drum is supported by a journal ball bearing at one end, the other end running directly in the casing. The bearing does not lead a demanding existence, and is unlikely to warrant attention during the normal life of the engine. The same can be said of the plain end, and neither should require more than a cursory inspection for wear or damage.

4   The selector drum tracks, on the other hand, are subjected to fairly high loadings at times, and may begin to wear after high mileages have been covered. The grooves should be examined in conjunction with the selector fork pins with which they engage. The pins which engage with the drum tracks are integral on the layshaft gear forks, therefore, if they are worn the forks must be renewed. On the mainshaft fork the pin can be renewed separately. It is normal to find polished areas in the selector drum grooves where pressure has been exerted, but most of the wear will take place on the selector fork pins. It is, therefore, the forks which will require renewal in cases where wear is severe. If there are any signs of scuffing, or overheating (indicated by blueing of the fork ends), there is a mis-alignment which must be rectified. The selector fork ends must slide freely in their grooves, without side play. Examine the grooves, especially where they change direction. The selector drum OD should also be measured if a considerable mileage has been covered. The OD must not be below 37.90 mm (1.492 in).

5   If the machine has shown a tendency to jump out of gear, check the various components of the detent mechanism, renewing the springs as a precaution.

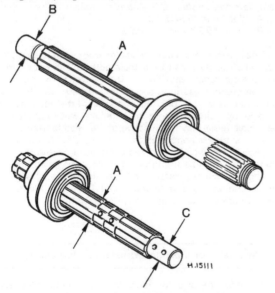

Fig. 1.16 Checking mainshaft and layshaft wear

36 Engine reassembly: general

1   Before reassembly of the engine/gear unit is commenced, the various component parts should be cleaned thoroughly and placed on a sheet of clean paper, close to the working area.

2   Make sure all traces of old gaskets have been removed and that the mating surfaces are clean and undamaged. One of the

best ways to remove old gasket cement is to apply a rag soaked in methylated spirit or where necessary, a gasket cement solvent. This softens the cement allowing it to be removed without resort to scraping and the consequent risk of damage.

3   Gather together all the necessary tools and have available an oil can filled with clean engine oil. Make sure all new gaskets and oil seals are to hand, also all replacement parts required. Nothing is more frustrating than having to stop in the middle of a reassembly sequence because a vital gasket or replacement has been overlooked.

4   Make sure that the reassembly area is clean and that there is adequate working space. Refer to the torque and clearance settings whenever they are given. Many of the smaller bolts are easily sheared if over-tightened. Always use the correct size screwdriver bit for the crosshead screws and never an ordinary screwdriver or punch. If the existing screws show evidence of mistreatment in the past, it is advisable to renew them as a complete set, using Allen screws in preference to cross-headed screws.

5   In addition to the above items, it will be necessary to obtain a tube of silicone rubber (RTV) jointing compound. This is used extensively in place of gaskets, particularly in the case of the crankcase joint. Also required is a can of molybdenum disulphide grease (Moly grease) for use as the initial lubricant whilst the main oil feed is circulating.

37 Engine and gearbox reassembly: refitting the gear selector internal mechanism

1   The right-hand end selector drum components may be assembled prior to insertion of the drum into the lower casing half, if required. Alternatively, the selector drum can be fitted and the external components fitted with the drum in position.

2   Fit the selector pin containing centre piece to the end of the selector drum. The innermost of the two plates which compose the centre piece, has one hole in its rear face. This must engage with the locating pin in the end of the drum. With the centre piece correctly fitted, offer up the outer index plate (the star-shaped plate). Note the locating hole in the rear face of the outer index plate, and the protruding pin on the outermost plate of the centre piece. This pin must locate with the index plate hole. Fit the central retaining bolt and tighten down firmly.

3   Slide the selector drum and its bearing part-way into the inverted upper crankcase half. Fit the largest of the three selector forks to the drum now, before the latter is pushed fully home. Ensure the fork is positioned correctly and secure it with the guide pin and retaining clip. If the fork is correctly positioned, the pin and clip will be facing the inner blind end of the drum. A pair of snipe-nose pliers will facilitate fitting of the pin and clip.

4   Fit the rear selector fork support shaft partway through the casing and place the two forks in position as the shaft is slid fully home. Ensure that the forks are refitted correctly, with the pins (which locate with the selector drum grooves) facing innermost on the shaft.

5   With the selector drum inserted correctly, ensure the left-hand end of the drum has not damaged the oil seal in the casing wall as it was passed through. Fit the neutral indicator switch securing it with the single crosshead screw, and then fit the switch operating cam to the end of the drum. Note that the cam has a tab which should locate with the slot in the end of the drum. Secure the switch cam with the washer and crosshead screw.

6   Position the positive stop lever/selector drum bearing retainer plate and secure it with the two hexagon-head bolts. Do not forget the small spring.

7   Assemble the longer neutral detent lever arm and the drum detent lever, collar and spring on the pivot bolt, but do not yet fit the assembly to the casing.

8   Lubricate the selector drum tracks with engine oil, then check the assembly for free operation. Check that the neutral switch contact is touching the switch contact tab.

53

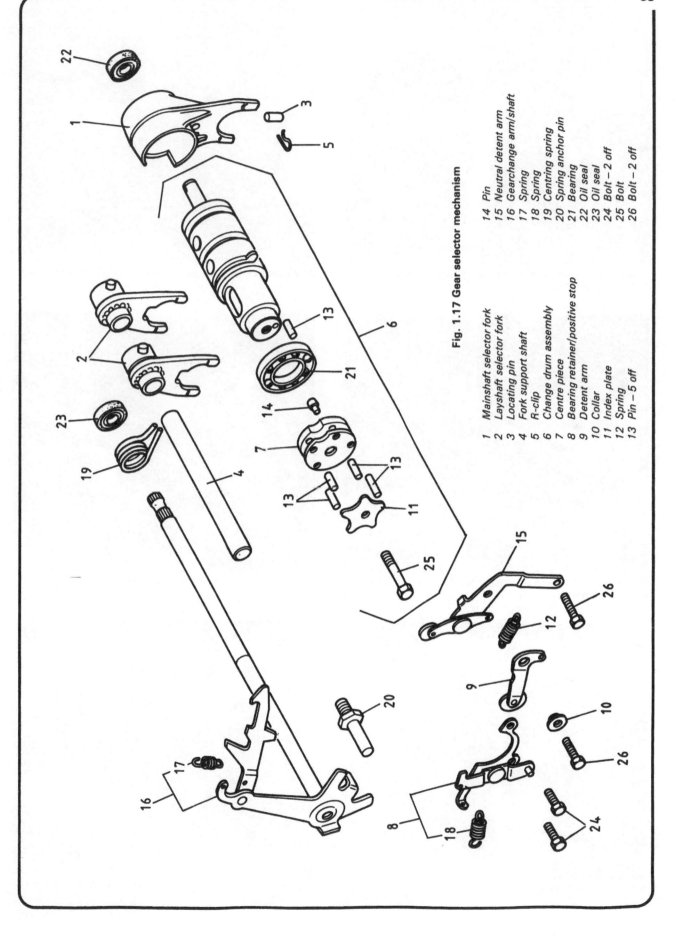

Fig. 1.17 Gear selector mechanism

1 Mainshaft selector fork
2 Layshaft selector fork
3 Locating pin
4 Fork support shaft
5 R-clip
6 Change drum assembly
7 Centre piece
8 Bearing retainer/positive stop
9 Detent arm
10 Collar
11 Index plate
12 Spring
13 Pin – 5 off
14 Pin
15 Neutral detent arm
16 Gearchange arm/shaft
17 Spring
18 Spring
19 Centring spring
20 Spring anchor pin
21 Bearing
22 Oil seal
23 Oil seal
24 Bolt – 2 off
25 Bolt
26 Bolt – 2 off

37.2a Fit centre piece to the end of the selector drum; note locating pin in drum end

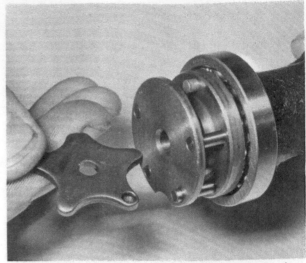

37.2b Locate the outer index plate with the pin on centre piece

37.3a Insert drum partially, then fit selector fork, and slide drum into correct position

37.3b Secure the fork with the guide pin and ...

37.3c ... the spring clip

37.4 Slide selector fork shaft through forks when positioned correctly

37.5a Fit the neutral indicator switch cam and ...

37.5b ... the switch unit

37.6 Do not forget spring when fitting stop lever/bearing retainer plate

the dowel pins in position in the left-hand mainshaft bearing boss and the right-hand layshaft bearing boss.
6   Ensure that the gear selection mechanism is in the neutral position. The two assembled shafts can now be fitted into the casing. Check that the half-rings, dowel pins and oil seal locate properly. As a check that the needle bearing races have correctly located with their dowel pins, the two horizontal line marks inscribed on the end of each bearing must be aligned with the jointing face of the casing. To help prevent oil leakage ensure that the layshaft oil seal is pushed inwards fully so that the sealing lip seats correctly on the bearing.
7   Spin the shafts to ensure that everything turns freely. Turn the selector drum to engage each gear in turn, then return to neutral. Liberally oil the gears and bearings.

39 Engine and gearbox reassembly: refitting the starter idler gear

1   Position the starter idler gear forward of the gearbox shaft assembly. Insert the idler gear shaft through the hole in the casing wall and locate the idler gear with the shaft.
2   Insert the bolt and tabbed locking plate to secure the idler gear shaft. Bend up the tabs on the locking plate to secure the retaining bolt.

38.1a Bearing and 1st gear pinion (14T) on mainshaft

38 Engine and gearbox reassembly: rebuilding and fitting the gearbox mainshaft and layshaft assemblies

1   The gearbox shafts should be reassembled by reversing the dismantling sequence. Reference should be made to the exploded view of the assembly and to the accompanying photographic sequence for details of the disposition of the various components. Although reassembly is generally straightforward, the following points should be noted.
2   Check that all the circlips seat squarely in their grooves, and that they are not slack when fitted. If in doubt, fit new circlips to preclude failure at a later date. Note the fitting of a tabbed locking washer and a splined washer between the layshaft 2nd gear and 3rd gear pinions. The locking washer must be twisted in order that it may engage correctly with the next splined washer.
4   Fit a half-ring in the bearing support boss groove of the right-hand end of the mainshaft. A half-ring should also be fitted to the inner groove of the left-hand layshaft bearing boss. Note that the outer groove is provided as a means of location for the oil seal. Turn the mainshaft left-hand needle roller bearing so that its dowel hole is facing downwards. Position the layshaft right-hand needle roller bearing in a similar manner.
5   Fit a new, greased, oil seal to the left-hand end of the layshaft. The closed side of the oil seal must face outwards. Fit

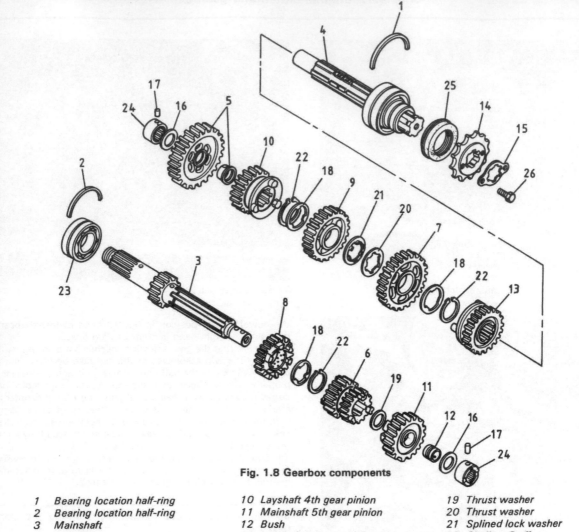

Fig. 1.8 Gearbox components

1	Bearing location half-ring	10	Layshaft 4th gear pinion	19	Thrust washer
2	Bearing location half-ring	11	Mainshaft 5th gear pinion	20	Thrust washer
3	Mainshaft	12	Bush	21	Splined lock washer
4	Layshaft	13	Layshaft 5th gear pinion	22	Circlip – 3 off
5	Layshaft 1st gear pinion	14	Final drive sprocket	23	Bearing
6	Mainshaft 2nd/3rd gear pinion	15	Sprocket retaining plate	24	Bearing – 2 off
7	Layshaft 2nd gear pinion	16	Thrust washer – 2 off	25	Oil seal
8	Mainshaft 4th gear pinion	17	Location dowel – 2 off	26	Bolt – 2 off
9	Layshaft 3rd gear pinion	18	Thrust washer – 3 off		

38.1b Fit 4th gear pinion (27T) and ...

38.1c ... secure with thrust washer and ...

38.1d ... circlip

38.1e Fit combined 2nd/3rd gear pinions (18T/21T) as shown

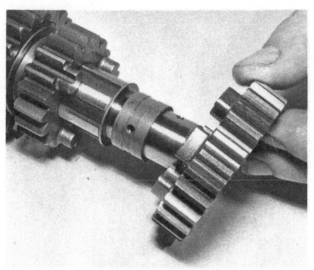

38.1f Position thrust washer, bush and 5th gear pinion (26T) on mainshaft

38.1g Fit thrust washer followed by needle roller bearing and cage

38.1h Fit heavy thrust washer to other end of mainshaft

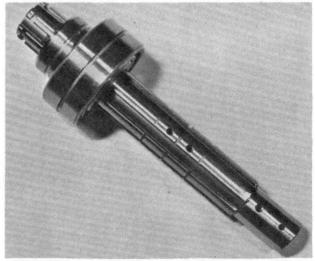

38.1i Layshaft and bearing ready for reassembly

38.1j Bearing is retained by this circlip

38.1k Fit layshaft 5th gear pinion (23T), dogs outermost, and ...

38.1l ... secure with circlip. Then ...

38.1m ... fit thrust washer and layshaft 2nd gear pinion (31T)

38.1n Position thrust washer followed by ...

38.1o ... a special tabbed locking washer

38.1p Slide the 3rd gear pinion (28T) into place ...

38.1q ... followed by thrust washer and ...

38.1r ... secure with circlip

38.1s Fit the 4th gear pinion (29T) ...

38.1t ... followed by the bush and large 1st gear pinion (35T)

38.1u Finish with the heavy thrust washer, needle roller bearing and cage

38.5 Fit dowel pin for left-hand mainshaft bearing and right-hand layshaft bearing

38.6a Position the mainshaft (note dowel pin and bearing half ring and oil seal groove)

38.6b Fit layshaft using new oil seal. Ensure selector forks engage with shafts

39.1 Position starter idler gear and insert shaft

39.2 Secure idler gear with locking plate and bolt

40 Engine and gearbox reassembly: refitting the crankshaft and primary shaft driven sprocket/starter motor clutch assembly

1   If the main bearing shells were removed, or have been renewed, check that they are all laid out in correct order and then refit them to their respective recesses. Ensure that the locating tang on each shell corresponds with the depression in which it engages. Ensure that each shell is firmly located before proceeding further. Apply a film of molybdenum disulphide grease to each of the main bearing surfaces.
2   Fit the camshaft chain and primary chain to their respective crankshaft sprockets, ensuring that the Hy-Vo chains are refitted in their original running directions. This, of course, does not apply if new chains are being fitted. Arrange the upper crankcase half on the workbench in the inverted position. The front section of the crankcase should be raised, using suitable wooden blocks, so that the connecting rods do not hinder crankshaft fitting by making contact with the top of the workbench.
3   Fit a new oil seal to each end of the crankshaft, having first lubricated their sealing lips with molybdenum disulphide grease.
4   Lower the crankshaft into position, ensuring that none of the bearing shells become displaced. Check that the oil seals

are located correctly in their grooves, and apply one or two drops of Loctite or a similar locking compound to retain the seals.

5   Engage the primary driven sprocket/starter clutch assembly with the rearward loop of the primary drive Hy-Vo chain and mesh the starter gears. Note that the correct fitting of this assembly is with the sprocket towards the alternator side of the crankcase.

41 Engine and gearbox reassembly: joining the crankcase halves

1   If the primary chain oil feed nozzle was removed for cleaning it should now be refitted. Fit a new O-ring below the oil feed nozzle's integral fixing plate prior to refitting the device.
2   Refit the two hollow locating dowels to the upper casing half mating surface.
3   Clean the crankcase mating faces with high flash-point solvent or methylated spirit to remove any residual grease spots or dust. Coat the lower casing half mating face with a silicone rubber (RTV) jointing compound, taking care not to obstruct any oilways.
4   **Important note:** On no account allow the jointing compound to get near the main bearing shells. A border of about $\frac{1}{8}$ in must be left clear to prevent the compound findings its way onto the bearing surface when the casing halves are joined. If this precaution is not observed the bearing oil feeds may become blocked, resulting in seizure. See the accompanying illustration.

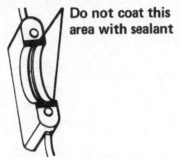

Do not coat this area with sealant

Fig. 1.19 Crankcase gasket compound application

5   Carefully lower the lower crankcase half onto the inverted upper half. Tap the lower half gently with a soft-headed mallet to seat fully the bearings and oil seals and make sure the joint closes all round the mating surfaces.
6   The majority of the crankcase holding bolts are fitted from the underside of the unit, and these should be dropped into position. The ten bolts which also retain the crankshaft assembly should each have a plain washer fitted. Note that the threads of these bolts should be lubricated with a smear of molybdenum disulphide grease. The above-mentioned bolts are 8 mm. The remainder are all 6 mm, except for one further 8 mm bolt (the longest of the crankcase holding bolts) fitted from the top, towards the rear of the crankcases. Do not forget the one 6 mm bolt fitted within the confines at the sump. Note the cable guide clip which should be fitted below the head of No 12 bolt.
7   Refer to the tightening sequence shown in the accompanying illustration, and tighten each bolt down to about $\frac{1}{2} - \frac{2}{3}$ of its recommended torque setting. Then go over each one again to bring it up to the full torque value. The torque settings are as shown below.

Crankcase torque settings
*   All 6 mm bolts: 1.0 – 1.4 kgf m (7 – 10 lbf ft)*
*   All 8 mm bolts: 2.2 – 2.6 kgf m (16 – 19 lbf ft)*

8   Turn the engine over, so it is the right way up, and fit the five 6 mm bolts and the one long 8 mm bolt. Then return the engine to its inverted position for the next Section.

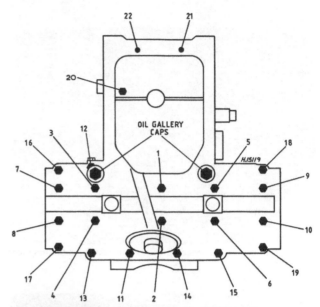

Fig. 1.20 Crankcase bolt tightening sequence

40.1 Liberally lubricate the main bearings

40.3 Lubricate new oil seal lips prior to fitting to crankshaft ends

40.4 Lower crankshaft assembly into position (both chains must be fitted)

40.5 Engage primary driven sprocket/starter clutch assembly with loop of primary chain

41.3 Use jointing compound carefully. Note hollow locating dowel

41.5 Lower the crankcase lower half into position

41.6 Do not forget this 6 mm bolt within confines of sump

41.7 Tighten all bolts to the recommended torque settings

42 Engine and gearbox reassembly: refitting the primary shaft, primary chain tensioner and external gear selector mechanism

1   If the primary chain tensioner has been removed for examination, it must be refitted at this stage. Place the tensioner assembly in position in the casing and fit its single retaining bolt and locking tab plate. Reconnect the tensioner spring.

2   Insert the primary shaft, together with the needle roller bearing and spacer, through the aperture in the lower crankcase wall. Push the shaft through the centre of the primary driven sprocket/starter clutch assembly. Several light taps with a soft-headed mallet may be required to drive the shaft fully home. Ensure that the bearings locate properly, and the needle rollers and spacer are not displaced from within the starter gear.

3   Fit the bearing retainer plate. This curved plate is secured by two hexagon-head bolts; install only the lower bolt at this stage.

4   Fit the heavy thrust washer and secure it with the circlip to the protruding end of the primary shaft.

5   Before refitting the gearchange lever operating shaft, check that the shaft oil seal in the casing wall is in good condition. If in any doubt, renew it as a precaution against leakage. The seal lips should be coated with grease prior to fitting the shaft. Take care not to damage the seal with the shaft splines during installation. It is worth wrapping the splines with PVC tape to protect the seal at this stage. Slide the shaft into position, lifting the claw mechanism over the end of the index plate. Check that the centring spring engages with its anchor pin. Temporarily refit the gearchange lever and check that all five gear positions plus neutral can be selected.

6   With the detent assembly ready to be fitted (refer to paragraph 7, Section 37) offer the assembly up to the casing. Insert and tighten the pivot bolt. The bolt which retains the bottom of the longer neutral detent lever arm occupies the remaining primary shaft bearing retainer thread. The bolt thus serves to retain the top of the bearing retainer plate and the bottom of the detent arm. Check that the detent lever roller engages with the cam profile on the outer index plate and that it is under spring pressure.

7   Fit the primary drive pinion and secure it with a circlip.

42.1 Secure primary chain tensioner with locking plate and bolt. Reconnect spring

42.2 Insert primary shaft; do not displace starter gear needle roller bearing or spacer

42.3 Secure bearing retainer plate with lower bolt only at this stage

42.5a Slide gearchange operating shaft and lever into place ...

42.5b ... and engage centring spring with anchor pin

42.6 Fit and secure detent assembly. Fit second retainer plate bolt

42.7a Fit primary drive pinion and ...

42.7b ... secure it with circlip

43 Engine and gearbox reassembly: refitting the oil strainer and sump

1   Refit the oil strainer screen to the central bore in the base of the lower crankcase. Note the two O-rings on the strainer spigot; these should be renewed if their condition is suspect. The integral clip on the strainer fits over the crankcase rib to secure the device.

2   Clean the sump (oil pan) gasket face, taking care not to scratch the surface. Position the original sump gasket or, if that is damaged in any way, a new gasket, then offer up the sump itself. Insert and tighten the ten 6 mm retaining bolts in a diagonal sequence to avoid any risk of distortion.

3   Insert and tighten the oil drain plug and its sealing washer, ensuring that the plug is not overtightened. With frequent recommended oil changes the thread can become worn. Too much force will strip the thread.

44 Engine and gearbox reassembly: refitting the oil pump

1   Refit the oil pump to its cavity in the left-hand crankcase wall. Ensure that the two large hollow dowels and their O-rings are refitted to the rear of the pump. Similarly, refit the small O-

ring, at the top of the rear of the pump, which fits over the protruding end of the oil control jet. Also make sure the large pump body O-ring is re-installed. Note that if any of these O-rings are suspect, they must be renewed.

2   Check that the small oil control jet is installed correctly before fitting the oil pump. The nose of the jet should be pointing away from the casing towards the oil pump.

3   Install the oil pump and secure it with the three cross-head retaining screws. Make sure that the pump body O-ring sealing face and the pump orifices are kept absolutely clean.

45 Engine and gearbox reassembly: refitting the clutch

1   Turn the engine over so that it is resting the right way up. Make sure the cam chain does not fall back into the depths of the crankcase as the engine is raised upright. Support the forward underside of the crankcase with suitable wooden blocks to present a level engine on which to continue re-assembly.

2   Place the plain thrust washer over the protruding end of the gearbox mainshaft. Fit the shouldered spacer, with the shouldered face outermost, followed by the clutch outer drum. Mesh the teeth on the clutch drums driven gear and the primary drive pinion.

3   If the clutch pressure plate, centre and clutch plates are still assembled, they can now be fitted as a unit. Alternatively, proceed as follows.

4   Note: lubricate all the clutch friction and plain plates with engine oil prior to reassembly.

5   Fit the clutch pressure plate, followed by the inner-most friction plate, then a plain plate and so on, alternately fitting a friction plate and a plain plate. Note that the friction plate denoted plate B in Fig. 1.3, is thicker than the remainder of the friction plates, and must be fitted last, in the outermost position.

6   Position the shock absorber device outboard of the thinner friction plate. Note that the two rings which comprise this device (the shock absorbing spring and its seat) are of differing thicknesses. The thinner of the two rings should be fitted first, with its cupped side facing inwards. If difficulty is encountered installing the device in this way, the spring ring and its seat may be fitted to the clutch centre prior to the installation of the latter.

7   Align the internal splines of the plain plates and fit the clutch centre, rotating the clutch centre as necessary to align the splines.

8   Place the Belville washer in position and then fit the clutch centre securing nut finger tight. Note the wording OUTSIDE on the Belville washer and fit accordingly.

9   Place the four clutch springs in position and offer up the

clutch release plate. Note that the dished face of the release plate must face inwards. Fit the bolts in a diagonal sequence, tightening each one in turn a little at a time.

10 Using the method adopted during dismantling (refer to Section 12), lock the clutch centre and tighten the slotted centre nut to a torque setting of 4.7 – 5.3 kgf m (34 – 38 lbf ft).

11 Refit the clutch release bearing retainer, the release bearing, and the pushrod to the centre of the release plate.

46 Engine and gearbox reassembly: refitting the clutch cover

1   The clutch/primary drive cover can now be refitted. If the clutch release mechanism components were removed from the cover, or new components were necessary, they can now be fitted to the cover. Grease all components upon assembly in the clutch cover.

2   Position the two hollow locating dowels in the casing wall, fit a new cover gasket, and push the cover into position. Fit the ten 6 mm hexagon-head bolts that secure the cover. Do not omit the three cable guide clamps which should be fitted below and to the rear of the cover. Tighten the bolts down evenly, in a diagonal sequence, to prevent distortion.

43.1 Note the two O-rings on the oil strainer spigot

43.2 Ensure the sump seal is undamaged or renew before refitting sump

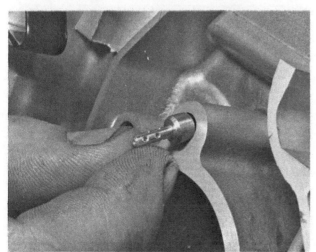

44.2 Small oil control jet must be free from blockage

44.3 Refit oil pump; ensure all O-rings and dowels are refitted

45.2a With the engine the right way up, ensure the heavy thrust washer is still fitted to mainshaft end, and ...

43.2b ... fit clutch outer drum and shouldered central spacer

45.5a Fit the pressure plate and innermost friction plate

45.5b Insert plain and friction plates alternately until ...

45.5c ... the outermost, thicker, friction plate is fitted

45.8 Fit the clutch centre and secure with Belville washer and slotted nut

45.9a Place the clutch springs in position and ...

45.9b ... fit the release plate. Tighten the centre nut fully before ...

45.10 ... fitting the release bearing and retainer and pushrod

46.2 Using a new gasket, refit the clutch cover

47 Engine and gearbox reassembly: refitting the automatic timing unit and pulser assembly

1   Offer up the automatic timing unit (ATU), ensuring that the locating pin at the rear of the unit fits into the hole in the shaft end. Fit the 6 mm securing nut to the end of the advancer shaft and, holding the crankshaft by fitting a spanner on the large hexagonal spacer provided, tighten the nut to a torque setting of 0.8 – 1.2 kgf m (6 – 7 lbf ft).
2   Note that the large hexagonal spacer can be used to hold or rotate the crankshaft as required, during future assembly or adjusting operations.
3   Position the CDI pulser assembly as it was prior to dismantling using the scribe marks made during dismantling as a guide. Fit and tighten the three crosshead securing screws through the holes in the periphery of the pulser mounting plate. Although use of the scribe marks has probably preserved accurate ignition timing, the timing should be checked soon after the engine is first started.
4   Feed the pulser assembly leads through the rubber grommet in the lower casing wall, and secure the leads beneath the guide clips below and to the rear of the clutch cover. Do not yet refit the circular inspection cover.

47.1 Fit the ATU and hexagonal spacer and secure with 6 mm nut

47.3 Refit CDI pulser assembly. Feed wires through grommet in casing lower wall

48 Engine and gearbox reassembly: refitting the pistons and cylinder block

1   Before replacing the pistons, pad the mouths of the crankcase with rag in order to prevent any displaced component from accidentally dropping into the crankcase.
2   Fit the pistons in their original order with the 'IN' mark on the piston crown towards the rear of the engine.
3   If the gudgeon pins are a tight fit, first warm the pistons to expand the metal. Oil the gudgeon pins and small-end bearing surfaces, also the piston bosses, before fitting the pistons.
4   Always use new circlips, **never** the originals. Always check that the circlips are located properly in their grooves in the piston boss. A displaced circlip will cause severe damage to the cylinder bore, and possibly an engine seizure.
5   Insert the two dowel pin/oil restrictors, one at each end of the crankcase mating surface, ensuring that the restrictors are not blocked. Fit new O-ring seals to each of these dowel pin/oil restrictors. Note that large diameter O-rings are fitted around each cylinder spigot where it protrudes from the cylinder block casting.
6   Place a new cylinder base gasket (dry) over the crankcase mouth.
7   Before fitting the cylinder block, install the forward cam chain guide blade in the casting.
8   Ensure that the piston rings are positioned so that their gaps are spaced approximately 120° out of line with each other, as shown in the accompanying illustration.
9   Arrange the pistons so that pistons No 2 and No 3 are at top dead centre (TDC).
10 Carefully lower the cylinder block whilst threading the camshaft chain up through the tunnel between the bores. This task is best achieved by using a piece of stiff wire to hook the chain through, and pull it up through the tunnel. The chain must engage with the crankshaft drive sprocket.
11 On an engine of this type, it is obviously an advantage to use piston ring compressors or clamps. Each cylinder bore, however, has a generous lead-in for the pistons at the bottom, and it is possible, in the absence of compressors, to gently lead the pistons into the bores compressing the rings at the same time, commencing with pistons No 2 and No 3. It is a distinct advantage here to have at least one other pair of hands to assist the operation. It is not practicable for the operator to hold the considerable weight of the cylinder block, and lower it gently, whilst leading the pistons in as well. Great care must be taken not to put too much pressure on the fitted piston rings.

12 With the two centre pistons engaged in their bores, remove the compressors if fitted, and continue to move the cylinder block downwards, together with the two centre pistons, so that the outer pistons will rise from bottom dead centre (BDC) as the crankshaft rotates. Engage these pistons as described previously. When all the pistons have finally engaged, remove the outer pistons compressors if fitted, remove the rag padding from the crankcase mouths and lower the cylinder block still further until it seats firmly on the base gasket.
13 Take care to anchor the camshaft chain through this operation to save the chain dropping down into the crankcase. The chain should be kept reasonably taut to prevent it from bunching around the crankshaft sprocket. This is particularly important when the crankshaft is turned, as the chain will tend to jam if left to its own devices. If required, there are two Honda service tools which can be usefully employed during this stage of reassembly; a pair of piston support blocks (07958 – 2500001) and a pair of piston ring compressors (07954 – 3740000).

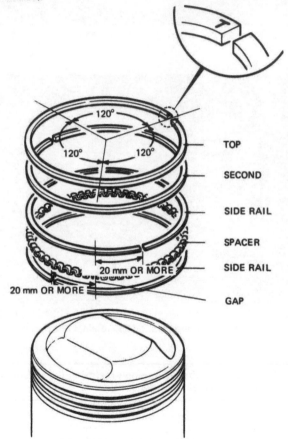

Fig. 1.21 Piston ring positioning

49 Engine and gearbox reassembly: refitting the alternator

1   Check that the crankshaft taper and the corresponding tapered bore of the alternator rotor are free from contamination, then place the rotor in position. It will be necessary to prevent crankshaft rotation while the rotor retaining bolt is tightened, using the same method employed during removal. Fit the headed retaining bolt and tighten to a torque setting of 5.0 – 6.0 kgf m (36 – 43 lbf ft).
2   The outer cover carries the stator and brushes, and can now be refitted. Check that the brushes are in a serviceable condition prior to installation, referring to Chapter 6 for details. When the cover is in position, route the output lead through the guides on the crankcase.

48.3 Oil gudgeon pin and small-end eye when refitting

48.4 Ensure the new circlips are correctly fitted in groove

48.5a Install the dowel pin/oil restrictors and ...

48.5b ... fit new O-rings, at each end of crankcase mating surface

48.6 Fit new base gasket. Note positioning of No 2 and No 3 pistons

48.10 With cam chain guide blade installed and chain retained by wire hook ...

48.12 ... engage the central pistons, followed by the two outer pistons

49.2 Refit the outer cover containing the alternator stator and brushes

50 Engine and gearbox reassembly: refitting the camchain tensioner and cylinder head

1   Insert the cam chain tensioner into the tunnel in the cylinder block so that the adjuster stud passes through the hole in the rear of the block. Ensure that the tensioner lower end locates correctly in the groove provided in the crankcase, as shown in the accompanying illustration. To aid subsequent camshaft installation, pull up on the top of the tensioner so that the blade flattens out into the retracted position. Lock the tensioner in this position by fitting the washer and domed lock nut. Fit the forward mounted chain guide blade, again checking that it is located correctly.

2   Check that the cylinder head and block mating surfaces are quite clean and free from all traces of old gasket material. Fit the two dowel pins and O-rings and then install the new cylinder head gasket.The gasket must be fitted with the wider turned edges of the individual cylinder grommets facing upwards. Lower the cylinder head into position, taking care to ensure that the camshaft chain is fed up through its central aperture. Check that the cylinder head seats squarely and that the dowels have located properly. If necessary, tap the top of the cylinder head using a soft-faced mallet to help seat it.

3   Fit the 6 mm cylinder head retaining bolts and the two longer 8 mm bolts. All the bolts are fitted with plain washers. Care should be exercised when securing the cylinder head, because it can easily become warped if uneven pressure is applied. The bolts should be tightened in a diagonal sequence, working from the centre outwards as shown in the accompanying illustration. Start by tightening the bolts to approximately half the final torque value, then go through the sequence once more to bring the bolts up to full pressure. The recommended torque settings are given below:

Cylinder head retaining bolts
6 mm bolts 1.0 – 1.4 kgf m (7 – 10 lbf ft)
8 mm bolts 2.4 – 3.0 kgf m (17 – 22 lbf ft)

4   Fit and tighten the two 'short 6 mm bolts which are located to the front and rear of the cam chain tunnel.

5   Fit the cam chain tensioner upper securing bolt and washer through the hole in the rear of the cam chain tunnel.

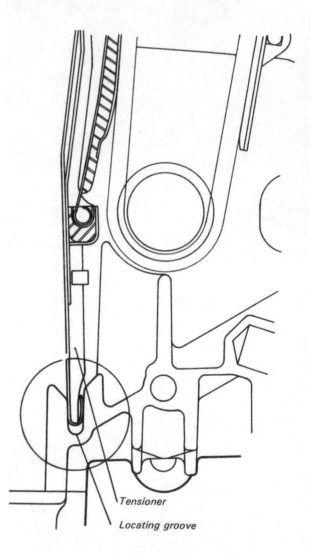

Tensioner

Locating groove

Fig. 1.22 Cam chain tensioner lower location

50.1 Insert cam chain tensioner blade; secure with washer and domed nut

50.2a With jointing face clean, install dowels and small O-rings and ...

50.2b ... make sure cylinder head gasket is fitted correctly

50.2c Lower cylinder head into position feeding cam chain up through tunnel

50.3 Tighten the cylinder head bolts evenly to the recommended torque settings

50.4 Do not forget the 6 mm bolts at front and rear of cam chain tunnel

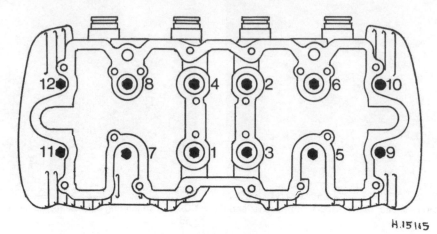

Fig. 1.23 Cylinder head bolt tightening sequence

51 Engine and gearbox reassembly: refitting the camshaft and setting the valve timing

1   Lubricate the cylinder head camshaft bearing surfaces with molybdenum disulphide grease.
2   Holding the camshaft chain taut to prevent it from bunching around the crankshaft sprocket, turn the crankshaft by means of the large hexagon spacer on the ATU until the '1,4T' mark appears in the timing window. Line the timing mark up against the index pointer in the top aperture (below the upper screw) of the pulser assembly mounting plate.
3   Engage the camshaft sprocket under the looped cam chain. Hold the camshaft sprocket and cam chain together, and slide the camshaft through the centre of the sprocket from the right-hand side. Note that the camshaft must be positioned so that the notched index mark on its end is located at the right-hand end of the cylinder head and is facing forward. Lower the camshaft onto its bearing surfaces in the cylinder head.
4   Align the camshaft end notch so that its centre is horizontal with the cylinder head mating surface. At this stage, one of the camshaft sprocket mounting bolt holes should be accessible and in line with the camshaft's threaded hole. Fit a securing bolt and tighten it gently.
5   Turn the crankshaft through 180° so that the remaining sprocket mounting hole becomes accessible and then fit the second bolt.
6   It is imperative that the camshaft is fitted in the position described above, failure to do so will result in incorrect valve timing. Slacken the cam chain tensioner locknut whilst rotating the engine and then retighten the locknut. Very carefully recheck the valve timing. It is only too easy to get the timing one tooth out. Finally, tighten the cam sprocket bolts to a torque setting of 1.4 – 1.8 kgf m (10 – 13 lbf ft).
7   Install the six rubber sealing plugs above cylinder head retaining bolts Nos 1, 2, 3, 4, 6 and 8.

52 Engine and gearbox reassembly: refitting the cylinder head cover/rocker gear

1   Install the two hollow locating dowels in the rear edge of the cylinder head mating surface. Ensure the mating surface is clean.

2   Clean the recessed gasket face of the cylinder head cover and, if it is in good condition, refit the original rubber sealing gasket. If damaged or perished, replace the gasket with a new item.
3   Before fitting the assembled cylinder head cover and rocker gear, slacken off fully all the tappet adjusting screws.
4   Refit the cylinder head cover, tapping gently with a soft-headed mallet if necessary to seat the cover on the two dowels.
5   Install the twenty-two 6 mm cylinder head cover retaining bolts. Note that twelve of the bolts (indicated on the accompanying illustration) should be fitted with sealing washers. Referring to the same illustration, tighten the cover retaining bolts in the sequence shown, in five or six stages, down to a recommended torque setting of 0.8 – 1.2 kgf m (6 – 9 lbf ft).
6   Reset the valve clearances using the tappet adjusters, as described in Routine Maintenance.
7   Refit the central breather cover and its rubber sealing gasket. Refit the two individual outer covers, again not forgetting their sealing gaskets. Fit the six retaining bolts (two on each cover) and their rubber mounting washers.

53 Refitting the engine and gearbox unit into the frame

1   As mentioned during the engine removal sequence, the engine/gearbox unit is fairly unwieldy, requiring at least two people to coax it back into position. This is even more important during reassembly, as the unit must be offered up at the right angle, and then manoeuvred into position. Care must be taken not to damage the finish on the frame tubes, and it is worthwhile protecting these with rag or masking tape.
2   Needless to say, the unit should be fitted from the right-hand side of the frame, where the removable lower section permits clearance. Again, bear in mind that the engine/gearbox unit is both heavy and awkward to manoeuvre; do not take chances which might result in damage to the machine or the operator. Note that it is important that the oil filter and housing are left off until the engine is installed. It must also be remembered that the final drive chain and layshaft-mounted final drive sprocket must be correctly positioned prior to installing the engine.

3 It is recommended that a jack of some description prefer-ably a trolley jack, is used to facilitate installation. This will take the weight of the unit whilst the mounting bolts are refitted. Always use a wooden block or rags between the crankcase and the jack.

4 Fit the various mounting bolts and plates in position by following a reversal of the removal procedure (Section 5), and as shown in the accompanying photographs. Do not tighten them until they are all in place. **Do not** force the bolts into position as this may damage the threads. Use a wooden lever between the frame and casings, in order to lift the engine slightly and align bolt holes.

5 Bear in mind the various footrest/brake pedal arrangements featured on the different models, as described in the removal sequence.

6 Once in position, the various bolts and nuts should be tightened to the values specified below.

Torque settings:

8 mm bolt/nut	2.6 – 3.2 kgf m (19 – 23 lbf ft)
10 mm bolt/nut	3.0 – 4.0 kgf m (22 – 29 lbf ft)
12 mm bolt/nut	8.0 – 10.0 kgf m (58 – 72 lbf ft)

51.2 Align the '1.4 T' mark with fixed pointer as shown

51.3 Pass the camshaft through the sprocket and cam chain and ...

51.4a ... align the camshaft end notch centre line with mating surface

51.4b Fit and tighten cam sprocket bolts one at a time

51.7 Refit the six rubber sealing plugs

Notch in camshaft right-hand end

Cylinder head upper mating surface

H.15112

Fig. 1.24 Valve timing – camshaft position

Align the centreline of the camshaft end notch with the mating surface of the cylinder head.

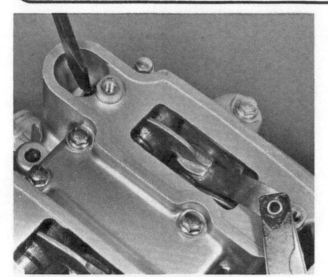

52.6 Reset the valve clearances to the correct settings

53.3 A jack positioned below the sump will facilitate engine refitting

53.4a Refit right-hand lower frame section and ...

53.4b ... fit forward lower mounting bolt and ...

53.4c ... the three upper bolts and bracket

53.4d Fit the long lower rear mounting bolt from the left-hand side and the two further bolts

53.4c Fit the long rear upper bolt with spacer, and ...

53.4f ... secure the bolt with the nut. Ensure battery earth lead is fitted behind nut

53.4g Secure front bracket and mounting bolts

53.5 Refit footrests/brake pedal mounting plate (UK CB650Z shown); note lower nut secures lower engine bolt

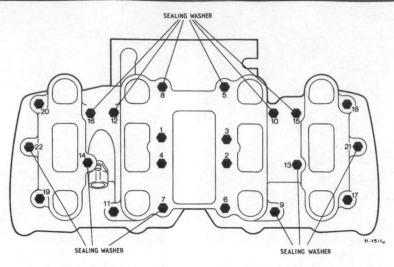

SEALING WASHER

Fig. 1.25 Cylinder head cover tightening sequence

54 Engine and gearbox unit installation: final assembly and adjustment

1 Reconnect the oil pressure switch lead to its terminal on the switch mounted on the top of the oil pump.

2 Replace the starter motor in its compartment in the upper crankcase behind the cylinder block. Ensure the O-ring on the splined inner end is in good condition prior to insertion of the motor. Refit and tighten the two 6 mm motor retaining bolts. Refit the chromed motor cover with its rubber sealing strip, and secure the cover with the two short 6 mm bolts.

3 Fit the final drive sprocket, together with the final drive chain, to the splined end of the layshaft. Refit the locking plate, by turning it until the splines engage, and secure the plate with the two 6 mm retaining bolts. Note that the rear wheel should be locked by applying the rear brake, whilst the bolts are being tightened. Do not omit to reset the drive chain tension by means of the rear wheel spindle adjusters.

4 Remake the connection from the alternator, ensuring that the leads are routed correctly and retained beneath the guide clips, at the block connector. Remake the connections from the starter motor, oil pressure switch and neutral indicator switch. Ensure the main earth lead was positioned below the nut of the upper rear mounting bolt upon its reinstallation.

5 Refit the left-hand side crankcase cover and secure it with the four retaining bolts.

6 If the sparking plugs have not yet been installed, they should now be fitted. Refit the sparking plug leads, noting that they are numbered 1 to 4, No 1 being the left-hand cylinder. Connect the tachometer drive cable at the cylinder head cover and secure it with the single locking bolt.

7 Whilst still relatively accessible, adjust the cam chain tension using the adjuster locknut at the base of the rear of the cylinder block. Place a spanner on the hexagonal spacer fitted to the ATU. The pulser assembly cover should not have been refitted as yet. Loosen the domed tensioner locknut. Using the spanner on the hexagonal spacer, slowly rotate the crankshaft clockwise (forwards). As the crankshaft is rotated, retighten the locknut. The tensioner will automatically set itself to the correct tension as the locknut is loosened.

8 Reassemble the exhaust system in the reverse order of that described for removal, using a new sealing ring in each exhaust port.

9 Place the forward air cleaner chamber loosely in position in the frame. On the CB650 A and CA models, (US market) the chromed chamber side panels can be refitted, if they were previously removed. Do not yet secure the air cleaner chamber rear to the front of the element-containing filter case. A certain amount of movement will be required to facilitate carburettor refitting.

10 Manoeuvre the bank of carburettors into position, pushing the air cleaner chamber rearwards slightly, to obtain sufficient clearance. Persuading the carburettors to engage in the rubber mounting stubs is not easy, and will demand a degree of patient manipulation with the aid of a small screwdriver. The operation is made easier if an assistant is available to deal with one side of the carburettor bank. Tighten the securing clips, then repeat the operation with the air cleaner chamber hose connections. When the carburettor bank is back in place, reposition the air filter chamber and tighten the large screw clip to secure the chamber to the element-containing case.

11 Reconnect the large diameter breather hose at the spigot on the breather cover on the cylinder head cover. Secure the hose with the spring clip.

12 Reconnect both the throttle cables at the operating quadrant and seat the cable outers against their respective stops. Adjust the throttle cable free play to give 2 – 6 mm measured at the flanged end of the throttle grip in relation to the adjacent switch housing. Also reconnect the choke cable.

13 Make a final check on the electrical system. Trace and reconnect the CDI pickup leads, alternator leads, etc. Reconnect the battery leads giving each terminal a coat of petroleum jelly to inhibit corrosion. Check that the electrical system functions properly, paying particular attention to the rear brake light switch, which may be in need of adjustment, and to the oil pressure and neutral light switches.

14 Refit the side panels, then slide the fuel tank into position, ensuring that the mounting rubbers at the front of the tank engage correctly. Secure the tank with the single bolt at the rear and refit the rubber support block. Reconnect the petrol feed pipe to the petrol tap and secure it with the spring clips. Check that the various drain and breather hoses are routed correctly.

15 Assemble the oil filter components in the filter housing, using a new element. The spring should be fitted into the housing first followed by the washer, filter element and then the spacer collar. Check that the O-rings on the filter housing and the housing bolt are in sound condition; renew the O-rings if their condition is suspect. Tighten the filter housing bolt to 2.7 – 3.3 kgf m (20 – 24 lbf ft). Check that the crankcase drain plug has been refitted and tightened. Fill the crankcase with 3.5 litres (6.2 Imp pint/3.7 US quart) of SAE 10W/40 engine oil, noting that the oil level must be checked and topped up after the engine has been started and run for a few minutes.

16 Reconnect the clutch cable to the actuating arm on the outer cover. Set the cable adjuster to give $\frac{1}{2}$ – $\frac{3}{4}$ in (10 – 20 mm) free play measured at the lever end.

17 Finally, check around the workbench area for any 'left-over' parts; these should be identified and refitted before attempting to start the newly-rebuilt engine.

54.1 Remake oil pressure switch and neutral switch connections; note guide clips for alternator cable

54.2a Ensure O-ring is in good condition and ...

54.2b ... refit the starter motor

54.2c Secure motor with retaining bolts and ...

54.2d ... refit chromed cover and sealing strip

54.3 Fit sprocket and chain together. Secure sprocket with plate and bolts

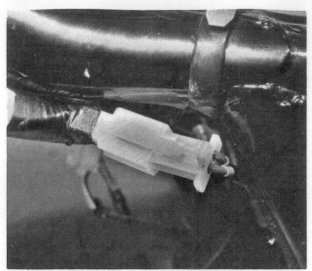
54.4a Remake connections for neutral and oil pressure warning lamps at this connector ...

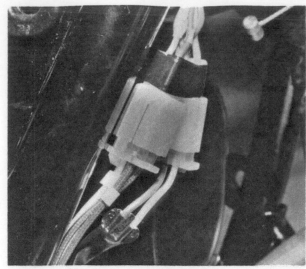
54.4b ... for CDI pulser generator here and ...

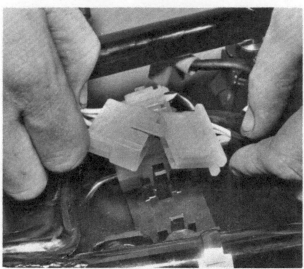
54.4c ... reconnect large connector carrying alternator wires

54.5 Refit left-hand crankcase cover

54.6 Secure tachometer cable with retaining bolt

54.8a Fit a new exhaust port gasket to achieve a gas-tight seal

54.8b Secure finned flanges with nuts on the mounting studs

54.9 Circular CDI inspection cover has no gasket

54.10 Position the air cleaner chamber before ...

54.11 ... manoeuvring the carburettors into the restricted space

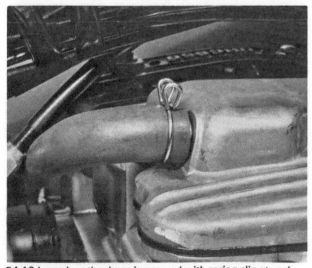

54.12 Large breather hose is secured with spring clip at each end

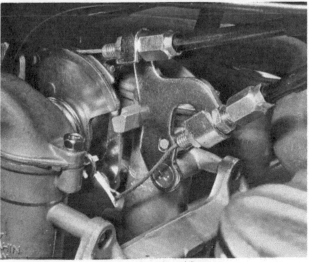

54.13 Refit and adjust both throttle cables

54.16a Install a new filter element and refit the filter housing

54.16b Tighten the filter housing bolt. **Do not** overtighten the bolt

54.16c Fill crankcase with correct quantity of recommended type of oil

55 Starting and running the rebuilt engine unit

1   Make sure that all the components are connected correctly. The electrical connectors can only be fitted one way, as the wires are coloured individually. Make sure all the control cables are adjusted correctly. Check that the fuse is in the fuse holder and try all the light switches and turn on the ignition switch. Close the choke lever to start.
2   Switch on the ignition and start the engine by turning it over a few times with the electric starter, bearing in mind that the fuel has to work through the four carburettors. Once the engine starts, run at a fairly brisk tick-over speed to enable the oil to work up to the camshaft, and valves. Check that the oil pressure warning lamp goes out after a few moments running.
3   The ignition timing must be checked and if necessary adjusted as detailed in Chapter 3, Section 7. The use of a stroboscopic lamp is recommended. Having accomplished this, refit the CDI pulser inspection cover.

4   Before taking the machine on the road, check that the brakes are correctly adjusted, with the required level of hydraulic fluid in the handlebar master cylinder.
5   Make sure the rear chain is correctly tensioned to $\frac{3}{4}$ – 1 inch (15 – 25 mm) up and down play.
6   Check the exterior of the engine for signs of oil leaks or blowing gaskets. Before taking the machine on the road for the first time, check that all nuts and bolts are tight and nothing has been omitted during the reassembling sequence.

56 Taking the rebuilt machine on the road

1   Any rebuilt engine will take time to settle down, even if the parts have been replaced in their original order. For this reason it is highly advisable to treat the machine gently for the first few miles, so that the oil circulates properly and any new parts have a reasonable chance to bed down.
2   Even greater care is needed if the engine has been rebored or if a new crankshaft and main bearings have been fitted. In the case of a rebore the engine will have to be run-in again as if the machine were new. This means much more use of the gearbox and a restraining hand on the throttle until at least 500 miles have been covered. There is not much point in keeping to a set speed limit; the main consideration is to keep a light load on the engine and to gradually work up the performance until the 500 mile mark is reached. As a general guide, it is inadvisable to exceed 4000 rpm during the first 500 miles and 5000 rpm for the next 500 miles. These periods are the same as for a rebored engine or one fitted with a new crankshaft. Experience is the best guide since it is easy to tell when the engine is running freely.
3   If at any time the oil feed shows signs of failure, stop the engine immediately and investigate the cause. If the engine is run without oil even for a short period, irreparable engine damage is inevitable.
4   When the engine has cooled down completely after the initial run, re-check the various settings, especially the valve clearances. During the run most of the engine components will have settled into their normal working locations.

57 Fault diagnosis: engine

Symptom	Cause	Remedy
Engine will not start	Defective sparking plugs	Remove the plugs and lay them on the cylinder head Check whether spark occurs when ignition is on and engine rotated
	Faulty ignition system	See Chapter 3
Engine runs unevenly	Ignition or fuel system fault	Check each system independently, as though engine will not start
	Blowing cylinder head gasket	Leak should be evident from oil leakage where gas escapes
	Incorrect ignition timing	Check accuracy and reset if necessary
Lack of power	Fault in fuel system or incorrect ignition timing	Check fuel lines of float chambers for sediment Reset ignition timing
Heavy oil consumption	Cylinder block in need of rebore	Check bore wear, rebore and fit oversize pistons if required
Excessive mechanical noise	Worn cylinder block (piston slap) Worn camshaft drive chain (rattle) Worn big-end bearings (knock) Worn main bearings (rumble)	Rebore and fit oversize piston Adjust tensioner or replace chain Fit replacement bearing Fit new bearings and seals
Engine overheats and fades	Lubrication failure	Stop engine and check whether internal parts are receiving oil. Check oil level in crankcase

58 Fault diagnosis: clutch

Symptom	Cause	Remedy
Engine speed increases as shown by tachometer but machine does not respond	Clutch slip	Check clutch adjustment for free play, at handlebar lever; check thickness of inserted plates
Difficulty in engaging gears, gear changes jerky and machine creeps forward when clutch is withdrawn Difficulty in selecting neutral	Clutch drag	Check clutch for too much free-play. Check plates for burrs on tongues or drum for indentations. Dress with file if damage not too great
Clutch operation stiff	Damaged, trapped or frayed control cable	Check cable and renew if necessary. Make sure cable is lubricated and has no sharp bends

59 Fault diagnosis: gearbox

Symptom	Cause	Remedy
Difficulty in engaging gears	Selector forks bent Gear clusters not assembled correctly	Replace with new forks Check gear cluster for arrangement and position of thrust washer
Machine jumps out of gear	Worn dogs on the ends of gear pinions	Renew worn pinions
Gear change lever does not return to original position	Broken return spring	Renew spring

Chapter 2 Fuel system and lubrication

For information relating to 1981-on models refer to Chapter 7

Contents

Specifications

	CB650 Z	CB650 A and CA
Fuel tank		
Total capacity	18 litre (4.0/4.8 Imp/US gal)	13.5 litre (3.5 US gal)
Reserve capacity	3.5 litre (0.8/0.9 Imp/US gal)	3.5 litre (0.9 US gal)
Carburettors		
Make	Keihin	Keihin
Type:		
UK model	PD51 A	–
US models	PD50 A	PD50 B
Main jet	92	90
Float level	12.5 mm (0.50 in)	12.5 mm (0.50 in)
Pilot screw setting:		
CB650 Z (UK)	$1\frac{1}{2}$ turns out	–
CB650 Z (US)	–	$1\frac{5}{8}$ turns out
CB650 A and CA	–	$2\frac{1}{8}$ turns out
Idle speed	1050±100 rpm	1050±100 rpm
Fast idle speed	2000±700 rpm	2000±700 rpm
Venturi diameter	26 mm (1.02 in)	26 mm (1.02 in)
Lubrication system		
Type	Wet sump, high pressure	
Filter	Pleated paper element, gauze sump strainer	
Oil capacity:		
Dry	3.5 litre (6.2 Imp.pint/3.7 US quart)	
At oil/oil filter change	3.0 litre (5.2 Imp.pint/3.2 US quart)	
Nominal oil pressure	54 – 60 psi (3.8 – 4.2 kg/cm^2) at 3000 rpm/80°C (176°F)	
	Measured at oil pressure switch take off	
Oil pump:		
Type	Trochoid	
Delivery rate	21.3 – 26.3 litre (39.2 – 46.4 Imp pint/22.5 – 27.8 US quart) per minute at 3000 rpm	
Inner rotor/outer rotor clearance (max) ..	0.15 mm (0.006 in)	
Outer rotor/Body clearance (max)	0.35 mm (0.014 in)	
Pump end clearance (max)	0.1 mm (0.004 in)	

1 General description

The fuel system comprises of a steel petrol tank from which fuel is fed by gravity to the four Keihin carburettors. A single petrol tap is located beneath the petrol tank, on the left-hand side. The tap has three positions, giving a normal supply of petrol, an emergency reserve position, and an 'off' position. A gauze strainer is incorporated in the tap to trap any foreign matter which might otherwise block the carburettor jets. The petrol tank filler cap flap incorporates a lock operated by the ignition key.

The four carburettors are interconnected by a linkage to ensure synchronisation. The two throttle cables are connected to a pulley mounted between the two central instruments, the throttles being opened and closed positively.

For cold starting a cable operated choke mechanism is fitted. The single choke cable is connected to a pulley fitted between carburettors No 3 and 4, and is linked to all four carburettors by two shafts, so that the mixture can be temporarily richened. The choke mechanism is operated by a handlebar-mounted choke knob.

When the engine has started, the choke can be opened gradually as the engine warms up, until full air is accepted under normal running conditions.

The engine draws air in via a large capacity plastic chamber and a further moulded plastic case which contains the air filter element.

Lubrication is effected by the wet sump principle in which the reservoir of oil is contained within a sump formed at the bottom of the engine. The oil is shared by the engine, primary drive and transmission components. The oil pump is of the trochoid type and is driven from the left-hand end of the primary drive shaft.

Oil is supplied under pressure, via a full-flow oil filter with a replaceable paper element, to the crankshaft and to the overhead camshaft and rocker gear. A secondary flow passes to the primary drive shaft and to the gearbox. All surplus oil drains down to the sump and is then recirculated. The pump itself is protected by a gauze strainer in the base of the oil pick-up chamber.

2 Petrol tank: removal and replacement

1   The petrol tank is retained by two guide channels which locate with a circular rubber buffer on each side of the steering head, and a bolt passing through a lug at the rear of the tank and into the frame. The tank also has a rubber support block on which it rests at the rear.

2   The petrol tank can be removed from the machine without draining the petrol.

3   Raise the dualseat (CB650 Z) or remove the seat by unscrewing the two mounting bolts (CB650 C and CA). Turn off the petrol tap and detach the petrol pipe, after first releasing the wire retaining clips. Pinch the 'ears' of the spring clip together to release their grip on the feed pipe. Remove the mounting bolt at the rear of the tank, lift up the tank at the rear and then ease the unit backwards until the location cups leave the rubber buffers. Lift the tank upwards off the machine.

4   When replacing the fuel tank, lift at the rear and push down onto the front rubber buffers, then secure the bolt at the rear and reconnect the fuel lines. Take care not to trap control cables or stray wires between the tank and frame tubes. If the location cups are a tight fit on the rubber buffers, apply a small amount of washing-up detergent to the buffers to ease refitting.

3 Petrol tap and filter: removal, dismantling and replacement

1   It is not necessary to drain the petrol tank if it is under half full, as the tank can be laid on its side on a clean cloth or soft material (to protect the enamel), so that the petrol tap is uppermost. The petrol pipe should be removed before unscrewing the petrol tap. The tap is released by slackening the gland nut which secures it to its mounting stub on the underside of the tank. Remove the tap carefully with the gauze filter column attached.

2   Separate the gauze filter column from the tap and wash it out in clean petrol. When refitting the filter, note that the flat on its lower end must locate with the flat on the tap. Ensure the rubber washer is replaced in the union gland nut. No further maintenance is feasible, as parts are not available for the tap. In the event of malfunction it must be renewed. Do not forget that fuel leakage can be dangerous.

Fig. 2.1 Petrol tank

1   Petrol tank
2   Petrol tap
3   Filter column
4   Clip – 2 off
5   Cap lock
6   Sealing washer
7   Moulding – 2 off
8   Clip – 2 off
9   Mounting rubber – 2 off
10   Rear rubber seat
11   Filler cap
12   Seal
13   Badge
14   Badge
15   Label
16   Shouldered collar
17   Bolt
18   Countersunk screw – 4 off
19   Screw – 2 off
20   Petrol pipe

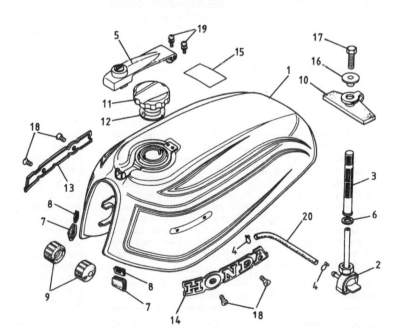

2.2a Petrol tank is retained by rubber buffers at forward end and ...

2.2b ... a single bolt and rubber support block at rear

3.1 Unscrew the gland nut to release tap and filter column

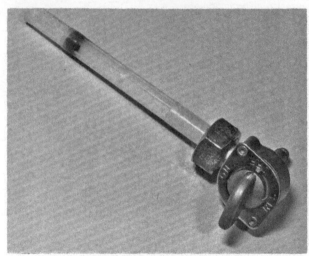

3.2 Gauze column may be cleaned. The tap cannot be dismantled

4 Petrol feed pipe: examination

1 A synthetic rubber feed pipe is used to convey petrol from the petrol tap to a T-pipe interconnected with the two left-hand carburettors. The pipe is retained by a wire clip at each end, which must hold the pipe firmly in position. Check periodically to ensure the pipe has not begun to split or crack and that the wire clips have not worn through the pipe surface.

2 **Do not** replace a broken pipe with one of natural rubber, even temporarily. Petrol causes natural rubber to swell very rapidly and disintegrate, with the result that minute particles of rubber would easily pass into the carburettors and block the internal passageways. Plastic pipe of the correct bore can be used as a temporary substitute, but it should be replaced with the correct type of tubing as soon as possible since it will not have the same degree of flexibility.

5 Carburettors: removal from the machine

1 To improve access to the carburettors it is suggested that the petrol tank is removed, as described in Section 2 of the Chapter, before dismantling proper commences.

2 Remove both side panels to provide access to the air cleaner chamber and air filter casing. Slacken the screw clips which secure the carburettors to the air cleaner chamber hoses, then slacken the large screw clip which secures the rear of the chamber to the front of the filter case followed by the filter case mounting bolts. The casing can now be pulled rearwards to provide a small clearance between the hoses and the carburettors.

3 Slacken the carburettor mounting clips. With the aid of an assistant, where possible, pull the assembly rearwards to free the mounting stubs from the inlet adaptors. There is little clearance available for this operation, and a degree of careful manoeuvring will prove necessary. Once clear of the stubs, withdraw the carburettors sufficiently to permit access to the throttle and choke cables. These should be disconnected by slackening off the adjusters and locknuts to allow the inner cables to be released. The complete bank of carburettors can now be pulled away from the machine. Note that there will be probably be slight spillage of residual fuel left in the float chambers as the carburettors are manhandled when detached. If required, this fuel can be drained off using the drain screws fitted to each float chamber.

5.3 Slacken the adjusters and disconnect throttle cables

6 Carburettors: separation and rejoining

1 Further dismantling is dependent on the degree of overhaul anticipated. Generally speaking, it is possible to attend to most areas of the instruments without separating them. Should complete dismantling prove necessary, bear in mind that a lot of work will be involved. Additionally, the manufacturer recommends that new choke valves, shafts and screws are fitted.
2 Start by unhooking the end of the choke relief spring from its mounting on the choke shaft of carburettors No 3 and 4. The spring is situated inboard of the No 3 carburettor.
3 Remove the tops from the four carburettors. Each top is retained by two crosshead screws. Remove the screw which passes through the forward end of each bellcrank and locates with the throttle link shaft.
4 Open the choke butterfly valve plates and carefully file off the staked ends of the securing screws, holding the carburettors so that the resulting metal filings fall out of the main bore. Remove and discard the securing screws, and lift the butterfly plates away.
5 The carburettors are mounted on a cast aluminium support bracket at the front, and two smaller brackets at the rear which also serve as cable anchor brackets. Release the front and rear mounting brackets. Note that the eight crosshead screws securing the front mounting bracket will probably be very tight, requiring the use of an impact driver to effect removal. Take great care not to damage the carburettors or the screw heads during this operation. The two rear brackets do not pose the same problem; each is retained by two hexagon-headed bolts.
6 Carefully separate the outer carburettors from the assembly. Take care not to damage the fuel and air connecting pipes or the mechanical linkage.
7 Slacken the locknut and loosen the adjuster screw on the mechanical throttle linkage positioned between carburettors No 2 and 3, where the throttle link rod passes through the linkage. Carefully separate the two carburettors with the same caution described in paragraph 6 above.
8 Reassembly is a direct reversal of the above sequences, noting the following points.
9 The central carburettors should be built up into a pair. Insert the throttle link shaft into the No 3 carburettor through the large throttle return spring. Do not yet fit the No 2 carburettor to the throttle shaft.
10 Ensure that the new O-rings are used on the various connecting pipes the T-pieces, lubricating each one with a smear of engine oil.
11 Install new choke link shafts in the outer facing sides of the central pair of carburettors.
12 Do not omit the nylon washer which fits between the No 2

carburettor body and the linkage. The central pair of carburettors can now be rejoined carefully so as not to damage the fuel or air pipes or throttle linkage.
13 The relevant hole in the throttle link shaft must be aligned with the throttle linkage set screw. When aligned, tighten the screw and secure it by tightening the locknut. Ensure the throttle linkage spring is refitted.
14 Carefully refit the outer carburettors to the assembled inner pair. Fit the front mounting bracket and install the retaining screws. Tighten the lower row from left to right, and then the upper row from left to right, in two or more stages to preserve carburettor alignment. At this stage, a check should be made to ensure the choke link shafts operate smoothly. If they do not, recheck the carburettor alignment. Fit the two rear mounting brackets and their securing bolts.
15 Align the relevant holes in the throttle link shafts and install and tighten the bell crank set screws.
16 Secure the large throttle return spring inner end to the rib on the No 3 carburettor body, and the outer end to the throttle link shaft arm.
17 Fit the choke butterfly valve plates and the new bolts and tab washers. **Do not** yet tighten the bolts.
18 Refit the choke relief spring to the choke link shaft arm between carburettors No 2 and 3. Operate the choke linkage and ensure it moves smoothly.
19 Tighten the choke butterfly plates retaining bolts and secure the bolts by bending up the locking tabs.
20 Check the operation of the linkages and levers, ensuring that they operate smoothly and do not bind or jam.

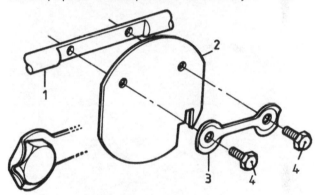

Fig. 2.2 Choke butterfly replacement

1	Control spindle	3	Tab washer
2	Butterfly valve	4	Screw – 2 off

6.2 Unhook choke relief spring from between carburettors No 3 and No 4

6.3 Remove the carburettor tops to gain access to bellcranks

6.4 Remove and discard butterfly plate screws and remove plates

6.5 Crosshead screws securing front mounting bracket will probably be tight

6.7a Slacken the locknut and the set screw as shown and ...

6.7b ... detach throttle linkage to allow ...

6.7c ... careful separation of the carburettors. Note the O-rings

6.13 Ensure throttle linkage spring is refitted

7 Carburettors: dismantling, examination and renovation

1   As mentioned previously, most of the normal overhaul jobs may be carried out with the instruments joined as a bank of four, thus avoiding a considerable amount of dismantling work. Note that where attention to the connecting linkages is required, it will be necessary to separate the instruments as described in Section 5.

2   Before any dismantling work takes place, drain out any residual fuel and clean the outside of the instruments thoroughly. It is essential that no debris finds it way inside the carburettors.

3   It is suggested that each carburettor is dismantled and reassembled separately, to avoid mixing up the components. The carburettors are handed and therefore components should not be interchanged.

4   Invert one carburettor and remove the three float chamber screws. Lift the float chamber from position and note the chamber sealing ring. This need not be disturbed unless it is damaged. The two floats, which are interconnected, can be lifted away after displacing the pivot pin. The float needle is attached to the float tang by a small clip. Detach the clip from the tang and store the needle in a safe place.

5   When unscrewing any jet, a close fitting screwdriver must be used to prevent damage to the slot in the soft jet material.

6   Unscrew the main jet whilst holding the jet needle holder with a small spanner. The holder may then be unscrewed to release the needle jet which is a push fit and projects into the carburettor venturi. Invert the carburettor to facilitate needle jet removal. Note that the slow jet, fitted beside the main jet column, is pressed into the carburettor body and cannot be removed. For this reason, when the jets are being cleaned, the slow jet can only be cleared by using compressed air in the jet passage.

7   Note that the pilot mixture adjuster screws on the 1980 US market CB650 models are fitted with adjustment limiter caps. These limiter caps cannot be detached from the pilot screws; if force is employed to try and remove the caps, the screws will be damaged.

8   Before removing the screws, screw them fully inwards until they seat lightly, recording the exact number of turns or part-turns necessary to seat them. Note that the pilot screws **must not** be tightened down against their seats as damage will be inflicted on the seats. By making a note of the exact position of each pilot screw prior to its removal, the need for resynchronisation upon reassembly will be negated. Unscrew and remove the pilot screws and their springs, plain washers and O-rings. Do not damage the screw threads when removing the plain washers and O-ring fitted to each screw.

9   A diaphragm air cut-off valve, is fitted to each carburettor to automatically richen the mixture on over-run, thus preventing backfiring in the exhaust system. The valve is located on the side of the main body, and thus will require separation of the instruments if attention is required. The cut-off valve is enclosed by a cover held on the outside of the carburettor body by two screws. Unscrew the screws holding the cover in place against the pressure of the diaphragm spring, and then lift the cover away. Remove the spring, and carefully lift the diaphragm and small O-ring.

10   An accelerator pump is fitted to the underside of carburettor No 2. If attention or examination of the device is required, remove the three pump cover retaining screws, detach the cover, and remove the spring and diaphragm.

11   Check the condition of the floats. If they are damaged in any way, they should be renewed. The float needle and needle seating will wear after lengthy service and should be inspected carefully. Wear usually takes the form of a ridge or groove, which will cause the float needle to seat imperfectly. If damage to the seat has occurred the carburettor body must be renewed because the seat is not supplied as a separate item.

12   After considerable service the throttle needle and the needle jet in which it slides will wear, resulting in an increase in petrol consumption. Wear is caused by the passage of petrol and the two components rubbing together. It is advisable to renew the jet periodically in conjunction with the throttle needle.

13   Inspect the cut-off valve diaphragm for signs of perishing or perforation. Damage will be easily seen.

14   Before the carburettors are reassembled, using the reversed dismantling procedure, each should be cleaned out thoroughly using compressed air. Avoid using a piece of rag since there is always risk of particles of lint obstructing the internal passage-ways or the jet orifices.

15   Never use a piece of wire or any pointed metal object to clear a blocked jet. It is only too easy to enlarge the jet under these circumstances and increase the rate of petrol consumption. If the compressed air is not available, a blast of air from a tyre pump will usually suffice.

16   Do not use excessive force when reassembling a carburettor because it is easy to shear a jet or some of the smaller screws. Furthermore, the carburettors are cast in a zinc-based alloy which itself does not have a high tensile strength. Take particular care when replacing the throttle valves to ensure the needles align with the jet seats. It should be noted that the throttle valve assembly which does not have a synchronisation adjuster should be fitted to the No 2 carburettor.

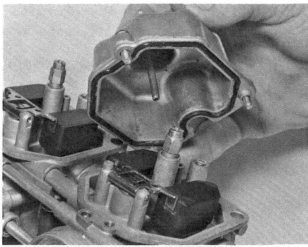

7.4a Remove float chamber noting the sealing ring

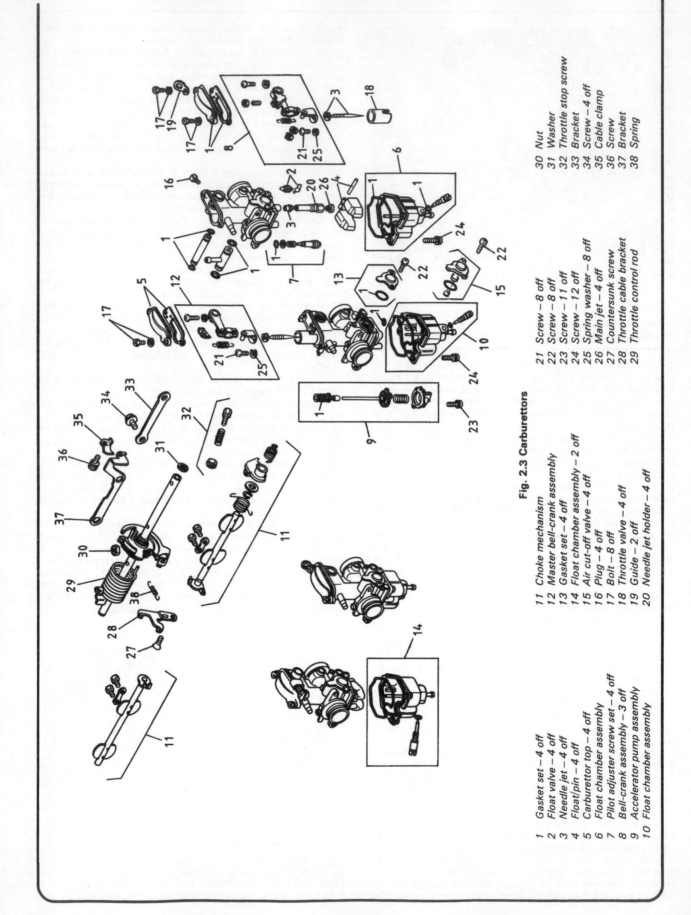

Fig. 2.3 Carburettors

1 Gasket set – 4 off
2 Float valve – 4 off
3 Needle jet – 4 off
4 Float/pin – 4 off
5 Carburettor top – 4 off
6 Float chamber assembly
7 Pilot adjuster screw set – 4 off
8 Bell-crank assembly – 3 off
9 Accelerator pump assembly
10 Float chamber assembly

11 Choke mechanism
12 Master bell-crank assembly
13 Gasket set – 4 off
14 Float chamber assembly – 2 off
15 Air cut-off valve – 4 off
16 Plug – 4 off
17 Bolt – 8 off
18 Throttle valve – 4 off
19 Guide – 2 off
20 Needle jet holder – 4 off

21 Screw – 8 off
22 Screw – 8 off
23 Screw – 11 off
24 Screw – 12 off
25 Spring washer – 8 off
26 Main jet – 4 off
27 Countersunk screw
28 Throttle cable bracket
29 Throttle control rod

30 Nut
31 Washer
32 Throttle stop screw
33 Bracket
34 Screw – 4 off
35 Cable clamp
36 Screw
37 Bracket
38 Spring

7.4b Displace the pivot pin and ...

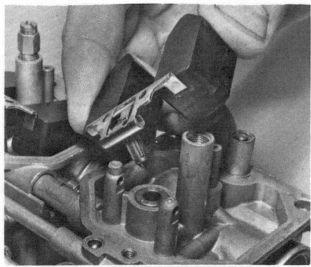

7.4c ... lift away the float assembly and small float needle

7.6 Slow jet (A) is a press fit and cannot be removed. Refer to text before touching pilot screw (B)

7.9 The air cut-off valve cover is held by two screws

7.10a Remove accelerator pump cover and spring, and ...

7.10b ... lift out the diaphragm

7.16a Refit the jet holder followed by ...

7.16b ... the main jet

8 Fast idle mechanism: adjustment

1 The fast idle mechanism takes the form of a pivoted arm mounted between the throttle and choke shafts on the No 2 carburettor. As the choke is operated, a cam on the choke shaft end bears upon the fast idle arm. Movement is transferred to the throttle stop, opening the throttle by a small amount to raise the cold start idle speed to 2000 ± 700 rpm.
2 Inspection and adjustment of the device must be carried out with the engine cold. Withdraw the choke operating knob fully, start the engine and check that the engine immediately runs at 2000 rpm ± 700 rpm. If the engine does not idle within this range, adjustment is required.
3 Stop the engine. Remove the petrol tank as described in Section 2 of this Chapter. Screw in the fast idle mechanism adjuster screw until it touches the cam on the choke link shaft. Push the choke operating knob fully inwards, and turn the adjuster screw inwards by $2\frac{1}{2}$ turns. Refit the petrol tank. Recheck the fast idle speed.

9 Accelerator pump: examination and adjustment

1 The CB650 models are equipped with an accelerator pump mounted on the underside of the No 2 carburettor. Its purpose is to richen the mixture during acceleration, thus allowing the carburettors to be jetted for a weaker overall mixture. This is ostensibly to enable the US market models to meet EPA emission regulations.
2 The pump is operated by a spring rod connected to the throttle cable quadrant. The rod is connected to a lever which terminates in a small metal tang. This depresses a rod to actuate the pump diaphragm. Fuel from the pump is fed by injection nozzles in each of the four instruments.
3 In the event of pump failure, the pump can be dismantled (see Section 7.9), and checked. Little can go wrong with the pump apart from a cracked or perished diaphragm, but if the pump system is completely dry it may require priming to expel air.
4 The pump output is set up during manufacture but should be re-checked after the carburettors have been overhauled or the operating linkage has been disturbed. With the throttle valve closed, check the clearance between the accelerator pump rod and the operating tang. This should be 0.4 – 0.6 mm (0.016 –

0.024 in). Any necessary adjustment may be made by careful bending of the operating tang.
5 At the other end of the arm, the pump stroke is limited by a second tang which stops against the projecting lug on the carburettor body. The specified gap here is 8 – 10 mm ($\frac{5}{16}$ – $\frac{3}{8}$ in). Once again, adjustment can be made by bending the tang.

10 Carburettors: synchronisation

1 For the best possible performance it is imperative that the carburettors are working in perfect harmony with each other. At any given throttle opening if the carburettors are not synchronised, not only will one cylinder be doing less work but it will also in effect have to be carried by the other cylinders. This effect will reduce the performance considerably. In the case of multi-cylinder engines especially, poor carburettor synchronisation will make the engine feel 'rough' and can induce backlash in the valve train or primary drive, thus giving the illusion of mechanical wear in these areas.
2 Synchronisation is carried out with the aid of vacuum gauges connected to the engine side of the carburettors. The gauges can take the form of mechanical clock-type instruments or a series of glass or plastic tubes, each containing a column of mercury. The latter type is often referred to as a mercury manometer. A suitable vacuum gauge set may be purchased from a Honda Service Agent, or from one of the many suppliers who advertise regularly in the motorcycle press.
3 Bear in mind that this equipment is not cheap, and unless the machine is regarded as a long-term purchase, or it is envisaged that similar multi-cylinder motorcycles are likely to follow it, it may be better to allow a Honda dealer to carry out the work. The cost can be reduced considerably if a vacuum gauge set is purchased jointly by a number of owners. As it will be used fairly infrequently this is probably a sound approach.
4 If the vacuum gauge set is available, proceed as follows. Remove the petrol tank so that access can be gained to the carburettors. Using a suitable length of feed pipe, reconnect the petrol tank with the carburettors, so that the petrol flow can be maintained. The petrol tank must be placed above the level of the carburettors. Remove the blanking plugs from the inlet tracts to the rear of the cylinder head, and connect the vacuum gauges to the engine.
5 Start the engine and allow it to run until normal working temperature has been reached. This should take 10 – 15 minutes. Set the throttle so that an engine speed of 1050 ±

100 rpm is maintained. If the readings on the vacuum gauge vary by more than 60 mm Hg (2.4 in Hg) it will be necessary to bring the carburettors within limits using the bell crank adjuster screws fitted to the top of carburettors No 1, 3 and 4. Note that if the readings on the gauges fluctuate wildly, it is likely that the gauges require heavier damping. Refer to the gauge manufacturer's instructions on setting up procedures.

6　The No 2 carburettor (second from left) is regarded as the base instrument: that is, it is non-adjustable and the remaining three carburettors must be adjusted to it. Honda produces a special combined screwdriver and socket spanner for dealing with the bell crank adjuster screws (Part number 07908 – 4220100). Its use makes the procedure easier, but it is not essential. Slacken the locknut of the adjuster concerned then turn the latter, noting the effect on the gauge reading. When the reading is as close as possible to that of the No 2 carburettor, hold the adjuster screw and retighten the locknut. Repeat the procedure on the remaining carburettors. Ensure all the locknuts are secure. Recheck the vacuum readings for a final check. Disconnect the vacuum gauges and refit the inlet tract take-off point blanking plugs. Refit and secure the carburettor tops. Refit the petrol tank.

11 Carburettors: idle adjustment

1　The engine idle speed should be checked and reset after the synchronizing operation has been carried out as described in the previous section. Before adjusting the carburettors a check should be made to ensure that the following settings are correct: ignition timing, valve clearance, sparking plug gaps, crankcase oil level. It is also important that the engine is at normal running temperature.

2　The pilot screws are fitted vertically in each carburettor, adjacent to the float bowl. These are set during assembly and should not be touched unless the carburettors have been overhauled or new pilot screws fitted. The master throttle stop screw is located at the centre of the carburettor bank and terminates in a knurled plastic knob. This should be set to give an idle speed of 1050 ± 100 rpm with the engine at normal temperature.

3　If consistent idling cannot be obtained by this method, and the carburettor synchronisation has been checked, it will be

necessary to check and adjust the individual pilot screw settings. To do this accurately, a test tachometer calibrated in 50 rpm increments will be required. The machine's tachometer is not sufficiently accurate for the test. Connect the test tachometer according to the manufacturer's instructions and allow the engine to reach normal operating temperature. The adjustment stages are detailed below.

4　Set each pilot screw to its nominal setting (CB650 with PD50A carburettor: $1\frac{5}{8}$ turns out; CB650 with PD50B carburettor: $2\frac{1}{8}$ turns out).

5　Set the master throttle stop control to give the prescribed idle speed of 1050 ± 100 rpm.

6　Unscrew each pilot screw by $\frac{1}{2}$ turn. If engine speed rises by 50 rpm or more, turn the screws out by another $\frac{1}{2}$ turn. Repeat until engine speed drops by 50 rpm or less.

7　Reset the idle speed using the master throttle stop control.

8　Turn the No 1 carburettor pilot screws inwards until the engine speed drops by 50 rpm, then back it out by $\frac{7}{8}$ of a turn for both types of carburettor.

9　Correct the idle speed once more.

10　Carry out the sequence detailed in paragraphs 8 and 9 for the remaining carburettors.

11　The above sequence is not detailed by the manufacturers for the PD51A carburettors as fitted to the UK market CB650 models. In the case of these models, it must suffice to set the pilot screws to their nominal $1\frac{1}{2}$ turns out. In most instances it will be sufficient to set the pilot screws on the US market CB650 models in the same way, assuming that a tachometer is not available. It follows that this simplified procedure will not produce such accurate results as those obtained by the method described earlier, and may prevent compliance with whatever local emission control regulations are in force.

12　Note that as stated in Section 7.7, there are small limiter caps fitted to the pilot screws on the 1980 US market CB650 models. If new pilot screws are fitted, new limiter caps must also be installed. It is not possible to detach the limiter caps from the existing pilot screws. When pilot screw adjustment has been completed as detailed above, the new limiter caps must be fitted. Use a small amount of Loctite, or a similar locking compound, to secure the caps to the screws. Note that the screws **must not** be moved as the limiter caps are fitted and secured. Honda recommend that new limiter caps should always be installed if new pilot screws have been fitted, in order to prevent maladjustment which could lead to a decrease in performance and an increase in exhaust emissions.

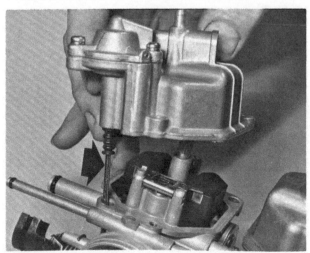

9.4 Accelerator pump is actuated by operating rod (arrowed)

10.5 Use the bell crank adjusters to synchronise the carburettors

11.2 Pilot screws should not be touched unless the carburettors have been overhauled

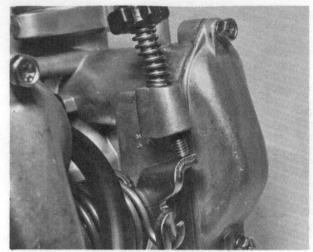

11.5 Set idle speed using the throttle stop screw

12 Carburettor settings

1   Some of the carburettor settings, such as the sizes of the needle jets, main jets and needle positions, etc, are pre-determined by the manufacturer. Under normal circumstances it is unlikely that these settings will require modification, even though there is provision made. If a change appears necessary, it can often be attributed to a developing engine fault.

2   Always err slightly on the side of a rich mixture, since a weak mixture will cause the engine to overheat. Reference to Chapter 3 will show how the condition of the sparking plugs can be interpreted with some experience as a reliable guide to carburettor mixture stength. Flat spots in the carburation can usually be traced to a defective timing advancer. If the advancer action is suspect, it can be detected by checking the ignition timing with a stroboscope.

13 Carburettors: adjusting float level

1   If problems are encountered with fuel overflowing from the float chambers, which cannot be traced to the float/needle assembly or if consistent fuel starvation is encountered, the fault will probably lie in maladjustment of the float level. It will be necessary to remove the float chamber bowl from each carburettor to check the float level.

2   If the float level is correct the distance between the uppermost edge of the floats and the flange of the mixing chamber body will be 12.5 mm (0.50 in).

3   Adjustments are made by bending the float assembly tang (tongue) which engages with the float tip, in the direction required (see accompanying diagram).

14 Exhaust system

1   Unlike a two-stroke, the exhaust system does not require such frequent attention because the exhaust gases are usually of a less oily nature.

2   Do not run the machine with the exhaust baffles removed, or with a quite different type of silencer fitted. The standard production silencers have been designed to give the best possible performance, whilst subduing the exhaust note to an acceptable level. Although a modified exhaust system, or one without baffles, may give the illusion of greater speed as a result of the changed exhaust note, the chances are that performance will have suffered accordingly.

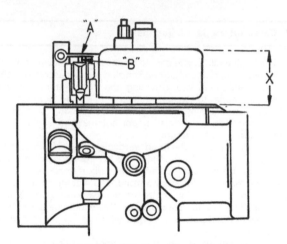

Fig. 2.4 Measuring the float height

15 Air filter: removing and cleaning the element

1   The air filtering job on the CB650 models is shared by two separate, though interconnected, casings. The forward air cleaner chamber is mounted immediately behind the carburettors into which the carburettor intakes fit. The rearward air filter case contains the removable filter element. The rear case is situated behind the left-hand side panel on CB650 Z models and behind the right-hand side panel on all later models.

2   The element can be removed for cleaning, or renewal as is necessary, after releasing the relevant side panel to reveal the access cover. The cover is retained by either two or three slot-headed bolts depending on the model. With the filter case cover removed, the filter element can be pulled out of the case.

3   The filter element is of the pleated paper type, and is supported by a metal mesh framework. The element can be cleaned by tapping it to dislodge any loose dust, and then blowing compressed air through from the inside. Check the paper surface for tears or contamination, renewing the element if it is excessively dirty or damaged.

4   On no account run the engine without the air cleaner attached, or with the element missing. The jetting of the carburettors takes into account the presence of the air cleaner and engine performance will be seriously affected if this balance is upset.

5   To replace the element, reverse the dismantling procedure. Ensure the filter case rubber seal is in good condition and is not omitted upon reassembly. Refit the filter case cover with the UP mark fitted accordingly.

6   Give a visual check to ensure that the inlet hoses are correctly located and not kinked, split or otherwise damaged. Check that the air cleaner case is free from splits or cracks.

7   Note the single drain tube from the forward air cleaner chamber. This has the function of disposing of oil and water which may accumulate within the chamber. Its lower end, which is normally closed to the atmosphere, should be squeezed occasionally to clear the accumulated fluid.

16   Engine lubrication

1   The engine shares a common lubrication system with the gearbox and primary transmission. Oil is picked up through a gauze strainer from the wet sump, and is drawn through the trochoidal oil pump. The pump is mounted externally on the left-hand side of the unit and is driven via gears from the primary shaft. Oil leaving the pump is controlled by a pressure relief valve which maintains a nominal pressure of $54 - 60$ psi ($3.8 - 4.2$ kg cm^2).

2   The oil splits into two routes at this junction, one of which directs the lubricant to the gearbox shafts where the bearings and pinions receive direct pressure lubrication. The oil enters the shaft centres from the left-hand end via a double seal arrangement, and exits via holes in the shafts to the pinion centres.

3   The main engine oil feed is conducted to the engine oil filter at the front of the crankcase where any small impurities are removed by the resin-impregnated filler element. The central filter bolt incorporates a bypass valve which operates in the event that the filter element becomes choked. This allows the oil to continue to circulate, albeit unfiltered.

4   After leaving the filter, the main oil supply is fed to the crankshaft main bearings via an oil gallery in the crankcase casting. Drillings in the crankshaft route the oil to the big-end bearings. Oil from the two outer crankshaft bearings is fed, via restrictor jets, to the camshaft bearing surfaces in the cylinder head. The emerging oil splashes serve to lubricate the pistons, cylinder walls and small-end bearings before running back to the sump.

5   Secondary feeds between the filter and oil gallery supply oil to the primary shaft and to the primary chain. An oil pressure switch is included in the lubrication circuit. It is mounted on top of the oil pump and operates a warning light in the event of an oil system failure.

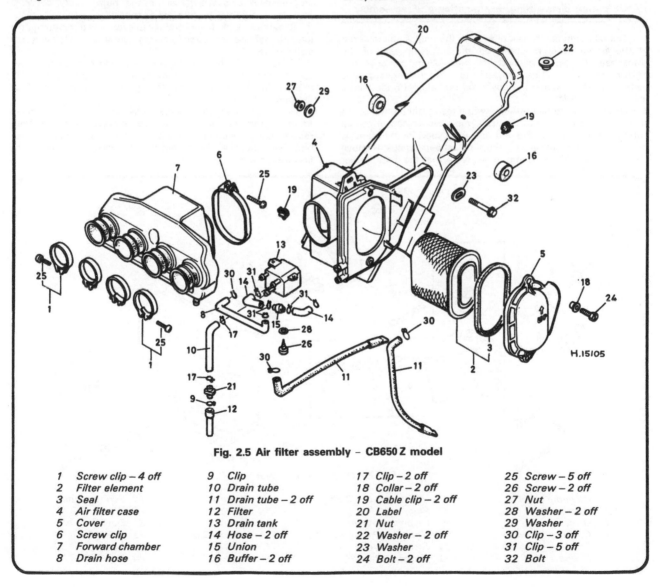

Fig. 2.5 Air filter assembly – CB650 Z model

1 Screw clip – 4 off	9 Clip	17 Clip – 2 off	25 Screw – 5 off
2 Filter element	10 Drain tube	18 Collar – 2 off	26 Screw – 2 off
3 Seal	11 Drain tube – 2 off	19 Cable clip – 2 off	27 Nut
4 Air filter case	12 Filter	20 Label	28 Washer – 2 off
5 Cover	13 Drain tank	21 Nut	29 Washer
6 Screw clip	14 Hose – 2 off	22 Washer – 2 off	30 Clip – 3 off
7 Forward chamber	15 Union	23 Washer	31 Clip – 5 off
8 Drain hose	16 Buffer – 2 off	24 Bolt – 2 off	32 Bolt

15.3 Air filter element simply pulls free after cover removed

15.5 Refit filter case cover correctly

17 Oil pump: dismantling, examination and reassembly

1   The oil pump can be removed with the engine unit in or out of the frame. It is mounted on the left-hand side of the crankcase, and can be reached after the left-hand rear cover has been removed. It will be necessary to remove the gearchange pedal prior to removing the cover, on models not fitted with a rear-set linkage.

2   Disconnect the oil pressure switch lead from the top of the switch. The pressure switch itself, mounted on top of the oil pump, can now be unscrewed and removed. The oil pump is retained to the crankcase by three crosshead screws. Remove the screws and lift the pump away. Note that the three further crosshead screws which secure the pump body cover in position, should not be touched at this stage.

3   Clean the pump body carefully to avoid any dirt entering the interior. Release the three securing screws and lift away the cover plate.

4   Displace the inner and outer rotor and remove the drive spindle and gear after withdrawing the small driving pin. Also remove and store the various O-rings and dowels used within the pump.

5   Wash all the pump components with petrol and allow them to dry before carrying out an examination. Before partially reassembling the pump for various measurements to be carried out, check the casing for breakage or fracture, or scoring on the inside perimeter.

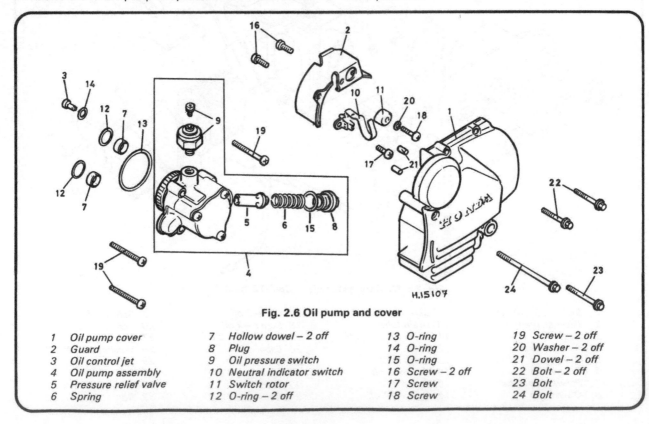

H.15107

Fig. 2.6 Oil pump and cover

1	Oil pump cover	7	Hollow dowel – 2 off	13	O-ring
2	Guard	8	Plug	14	O-ring
3	Oil control jet	9	Oil pressure switch	15	O-ring
4	Oil pump assembly	10	Neutral indicator switch	16	Screw – 2 off
5	Pressure relief valve	11	Switch rotor	17	Screw
6	Spring	12	O-ring – 2 off	18	Screw

19	Screw – 2 off
20	Washer – 2 off
21	Dowel – 2 off
22	Bolt – 2 off
23	Bolt
24	Bolt

6 Reassemble the pump rotors and measure the clearance between the outer rotor and the pump body, using a feeler gauge. If the measurement exceeds the service limit of 0.35 mm (0.014 in) the rotor or the body must be renewed, whichever is worn. Measure the clearance between the outer rotor and the inner rotor, using a feeler gauge. If the clearance exceeds 0.15 mm (0.006 in) the rotors must be renewed as a set. With the pump rotors installed in the pump body lay a straight edge across the mating surface of the pump body. Again with a feeler gauge measure the clearance between the rotor faces and the straight edge. If the clearance exceeds 0.10 mm (0.004 in) the rotors should be renewed as a set.

7 Examine the rotors and the pump body for signs of scoring, chipping or other surface damage which will occur if metallic particles find their way into the oil pump assembly. Renewal of the affected parts is the only remedy under these circumstances, bearing in mind that the rotors must always be renewed as a matched pair.

8 The pressure relief valve incorporated into the oil pump can be dismantled now, or it can be dismantled with the pump reassembled. Slacken and carefully remove the hexagon-headed relief valve sealing bolt. Note the O-ring on the inside of the bolt. Remove the valve spring and the valve plunger itself. Inspect the bore and plunger for wear, damage or contamination, cleaning or renewing the necessary parts as required.

During reassembly, lubricate the plunger with oil, having made sure that the valve components are perfectly clean.

9 Reassemble the pump components by reversing the dismantling procedure. The component parts must be ABSOLUTELY clean or damage to the pump will result. Replace the rotors and lubricate them thrroughly, before refitting the cover. Remember that the punch marked faces of the two rotors must both face upwards and should be aligned upon reassembly. Also align the notch in the inner rotor with the drive pin. Make a final check that everything is well lubricated, in particular that there is plenty of oil between the two rotors, before refitting the cover plate. Do not omit the large O-ring seal behind the cover plate.

10 Before refitting the oil pump to the engine, check the small oil restrictor jet which fits in the crankcase wall behind the pump. Examine the restrictor and ensure it is not blocked. Refit the restrictor in the correct way. Ensure its O-ring in the pump body is refitted.

11 Install the two hollow dowels and their O-rings, and the larger pump body O-ring seal, in position in the rear of the oil pump. Offer the pump up to the crankcase, so that the driven gear engages with the drive gear in the casing. Secure the pump with the three screws. Apply a small quantity of locking fluid to the thread on the oil pressure switch prior to its refitting. Reconnect the pressure switch lead. Refit the left-hand rear crankcase cover and the gearchange lever (if removed).

17.3 Lift away the pump cover plate and ...

17.4a ... displace the inner and ...

17.4b ... outer rotors

17.4c Displace the drive pin and ...

17.4d ... remove the drive gear and spindle

17.6a Check inner rotor to outer rotor clearance and ...

17.6b ... outer rotor to pump body clearance

17.6c Measure the end clearance with a straight edge and feeler gauge

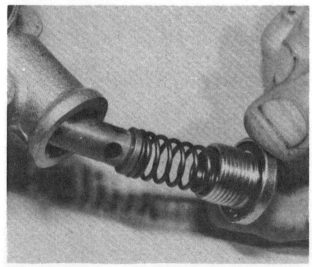
17.8a Remove pressure relief valve assembly noting the O-ring inside the cap bolt head

17.8b Inspect the assembly carefully. Lubricate before reassembly

17.9 Refit or replace large O-ring. Note both rotor punch marks

17.11a Install the two hollow dowels and ...

17.11b ... their O-rings

17.11c Do not omit the small O-ring for the restrictor jet

17.11d Refit and secure the cover plate

18 Oil strainer: location and cleaning

1   The oil strainer assembly is housed within the sump, and can be checked after the latter has been detached. Although not detailed in the routine maintenance schedule, it is worth cleaning the sump area occasionally because the nature of the sediment which collects in the sump will give an indication of the overall condition of the machine.

2   Remove the drain bolt and drain the contents of the sump. This is best done with the engine at its normal operating temperature to ensure full drainage. Remove the ten sump retaining bolts. Lift the sump away. The strainer screen is secured to a crankcase rib by a built-in clip on the strainer body.

3   Detach the gauze strainer and flush it clean with petrol or a low flash-point solvent where available. Take note of any metallic particles found, because these can often be linked to unusual noises that may have been noticed. Bronze-coloured particles will indicate that a bush is wearing or breaking up, whilst greyish particles may be the result of wear in the main or big-end bearings, or in the casings. Obviously, any large pieces of metal will warrant further investigation.

4   When refitting the strainer, note and examine the two O-rings on the strainer spigot where it enters the bore in the lower crankcase. Renew the O-rings if they are suspect in any way. When refitting the sump, use a new rubber sealing ring if necessary. Tighten the sump retaining bolts in a diagonal sequence to avoid unequal stress.

19 Oil filter: renewing the element

1   The oil filter is contained within a semi-isolated chamber at the front of the crankcase. Access to the element is made by unscrewing the filter cover centre bolt, which will bring with it the cover and also the element. Before removing the cover, place a receptacle beneath the engine, to catch the engine oil contained in the filter chamber.
2   When renewing the filter element it is wise to renew the filter cover O-ring at the same time. This will obviate the possibility of any oil leaks. Do not overtighten the centre bolt on replacement, the correct torque setting is 2.7 – 3.3 kgf m (20 – 24 lbf ft).
3   The filter by-pass valve, comprising a plunger and spring, is situated in the bore of the filter cover centre bolt. It is recommended that the by-pass valve be checked for free movement during every filter change. The spring and plunger are retained by a pin across the centre bolt. Knocking the pin out will allow the spring and plunger to be removed for cleaning.
4   When installing the filter components into the chamber fit the spring first followed by the washer, filter element and then the spacing collar.
5   Never run the engine without the filter element or increase the period between the recommended oil changes or oil filter changes. Engine oil should be changed every 4000 miles, as should the filter element. Use only the recommended viscosity of oil.

20 Oil pressure warning lamp

1   An oil pressure warning lamp is incorporated in the lubrication system to give immediate warning of excessively low oil pressure.
2   The oil pressure switch is screwed into the top of the oil pump on the left-hand side crankcase wall. The switch is interconnected with a warning light on the lighting panel on the handlebars. The light should be on whenever the ignition is on but will usually be out by about 1500 rpm, after the initial start-up.
3   If the oil warning lamp comes on whilst the machine is being ridden, the engine should be switched off immediately, otherwise there is a risk of severe engine damage due to lubrication failure. The fault must be located and rectified before the engine is re-started and run, even for a brief moment. Machines fitted with plain shell bearings rely on high oil pressure to maintain a thin oil film between the bearing surfaces. Failure of the oil pressure will cause the working surfaces to come into direct contact, causing overheating and eventual seizure.
4   If low oil pressure is experienced, it can be checked by connecting a suitable gauge to the switch take-off. At normal operating temperatures, ie approximately 80°C (176°F), a reading of 3.8 – 4.2 kg/cm^2 (54 – 60 psi) at 3000 rpm should be indicated. Any substantial drop in pressure should be investigated promptly if engine damage is to be avoided. Low oil pressure may be caused by very badly worn main bearings, oil pump leakage or pump failure. In rare cases the pressure relief valve sticking open may be found to be the culprit.

19.2 When renewing the element, check the filter housing O-ring

20.2 Oil pressure switch is screwed into the top of the oil pump

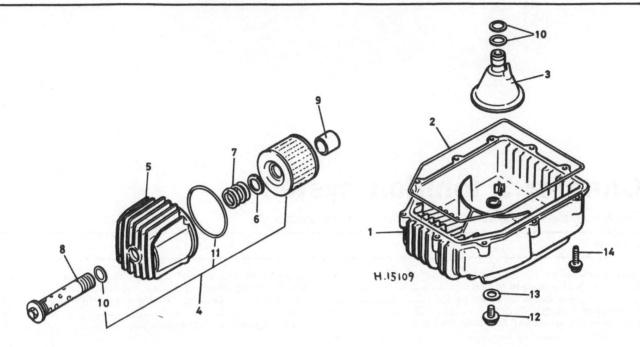

H.15109

Fig. 2.7 Oil filter and sump

1 Sump	6 Washer	11 O-ring
2 Sealing ring	7 Spring	12 Drain plug
3 Oil strainer	8 Bolt/by-pass valve	13 Sealing washer
4 Filter assembly	9 Spacer collar	14 Bolt – 10 off
5 Filter housing	10 O-ring	

21 Fault diagnosis: fuel system and lubrication

Symptom	Cause	Remedy
Engine gradually fades and stops	Fuel starvation Sediment in filter bowl or float chamber	Check vent hole in filler cap Dismantle and clean
Engine runs badly. Black smoke from exhausts	Carburettor flooding	Dismantle and clean carburettor. Check for punctured float or sticking float needle
Engine lacks response and overheats	Weak mixture Air cleaner disconnected or hose split Modified silencer has upset carburation	Check for partial block in carburettors Reconnect or renew hose Replace with original design
Oil pressure warning light comes on	Lubrication system failure	Stop engine immediately. Trace and rectify fault before re-starting
Engine gets noisy	Failure to change engine oil when recommended	Drain off old oil and refill with new oil of correct grade. Renew oil filter element

Chapter 3 Ignition system

For informaton relating to 1981-on models refer to Chapter 7

Contents

Specifications

Ignition system
Type .... Capacitor discharge ignition (CDI)

Ignition timing
Retarded .... 10° BTDC @ 950 – 1150 rpm
At full advance (2725 rpm) .... 28° 30' BTDC @ 2725 rpm

Sparking plugs
Make .... NGK or Nippon Denso (ND)
Type:
CB650 Z (UK) .... DR8ES-L or X24ESR-U
CB650 (all US models) .... D8EA/DR8ES-L or X24ES-U/X24ESR-U
Gap .... 0.6 – 0.7 mm (0.024 – 0.028 in)

Torque settings
Automatic timing unit (ATU) bolt .... 2.5 – 2.9 kgf m (18 – 21 lbf ft)
Sparking plugs .... 1.2 – 1.6 kgf m (9 – 12 lbf ft)

1 General description

The Honda CB650 models make use of a capacitor discharge ignition (CDI) system, in which the ignition timing and triggering is controlled electronically. As a result, the spark at the plugs is more powerful and is accurately timed. Because there are no mechanical aspects to the ignition system, wear does not take place, and thus the ignition timing will remain accurate unless disturbed, or in the rare event of component failure.

Like most four cylinder machines, these Honda models employ what is effectively two ignition systems, each controlling two cylinders. This means that each time a spark is generated in one of the ignition coils, it is fed to two of the sparking plugs. Combustion will only take place in the cylinder which is under compression, leaving one spark wasted. This arrangement is known as the 'spare spark' system. It is important to remember this when dealing with ignition system faults, as these are most likely to be confined to one half of the system. The system can therefore be divided into that which

related to cylinders 1 and 4, and the corresponding equipment relating to cylinders 2 and 3.

The system is triggered by a reluctor mounted on the crankshaft and which rotates past two pickup coils or pulsers. The function of the reluctor can be likened to that of the contact breaker cam in a conventional system. As the peak of the reluctor passes the pulser coil, it generates a small signal current. This current triggers the appropriate circuit in the ignition amplifier, or spark unit. The high voltage pulse from the amplifier feeds to the relevant coil, where its passage through the coil's primary windings induces the required high tension (HT) spark in the secondary windings. The reluctor continues to rotate, causing the same chain of events to take place in the remaining circuit.

For this range of models, Honda have chosen to retain a mechanical automatic timing unit (ATU) rather than opt for full electronic advance. As the engine speed rises, small weights on the ATU are thrown outwards against spring pressure. This movement is translated to the reluctor, which advances in relation to the crankshaft, thus providing an advanced spark for high speed running.

2 Electronic ignition system: testing

1 The CDI system, as mentioned previously, can be divided into two separate systems for most fault diagnosis purposes. Total failure of the entire system is unlikely, and trouble is usually confined to 1 and 4 cylinders or to 2 and 3 cylinders. In this case, it can be assumed that the system is functional up to the amplifier, or spark unit.

2 A preliminary check may be made as follows. Remove the sparking plugs from cylinders 1 and 2, remembering that these each represent one sub-system. Remove the inspection cover from the right-hand side of the unit to reveal the pulser coils and reluctor. Check that the engine kill switch and the main ignition switch are on.

3 Using a screwdriver with an insulated handle, bridge the gap between the reluctor and the metal core of one of the pulser coils. If the ignition circuitry is sound a spark will occur at the appropriate plugs. Repeat the check with the remaining pulser and plugs. If there is a defect in the system one of the plugs will fail to spark.

4 If no spark at all was found, attention should be turned to the ignition switch. Check that the switch operates properly, and that power is being fed to the ignition amplifier (spark unit). Ensure that the kill switch is at the run position and has not shorted. If all seems well, and the remaining wiring connections are sound, it can be assumed that the amplifier is in need of attention. As it is unlikely that both pulsers, both coils or all four sparking plugs will have failed simultaneously, they can be ignored for the time being.

5 If one of the plugs failed to produce a spark in the above test, the system can be assumed to be sound up to the amplifier unit. The latter should be checked, followed by the pulser coil(s) and ignition coil(s).

3 Ignition amplifier (spark unit): testing

1 If preliminary checks have indicated a possible fault in the amplifier unit it should be tested as described below. Note that a dc voltmeter, or a multimeter set on the appropriate scale, will be required for the test. It is assumed that owners with access to this equipment will be conversant with its use. If this is not the case, the work should be entrusted to a Honda Service Agent or an Auto-electrician.

Test A

2 Trace the pulser leads back to the white 4-pin connector block behind the right-hand side panel. Disconnect the pulser leads at the 4-pin connector block. Connect a multimeter set on 0 – 50 volts dc, as follows. Attach the positive (+) probe to the blue wire with yellow sleeve at the white 6-pin connector block. Connect the negative (-) probe of the multimeter to earth. Using a suitable length of wire as a jump lead, connect one end to the blue lead with the white tube at the wiring harness side of the 4-pin connector block. Switch on the ignition, and flash the end of the jump lead to earth. This operation should result in the meter needle fluctuating between 12V and 0V if the unit is working correctly.

Test B

3 Repeat the above test sequence but this time with the positive multimeter probe connected to the yellow lead of the other white 6-pin connector. Similarly, attach the jump lead to the yellow wire with the white tube at the 4-pin connector block. This will test the functioning of the amplifier unit for cylinders 2 and 3, whereas Test A tested the operation of the amplifier unit for cylinders 1 and 4. Similar results should be obtained in both tests. If the unit(s) prove faulty, they must be renewed since there is no practicable form of repair. Refer to the accompanying circuit diagram for details of the test connections.

3.2 The four connector blocks on 1979 models. Spark unit is at top right of photograph

4 Pulser coils: testing

1 The pulser coils can fail in two possible ways: either by an internal break in the windings (open circuit) or by an internal failure of the winding insulation (short circuit). A pulser in good condition will show a specific resistance reading when checked as described below.

2 Disconnect the pulser leads at the red connector behind the right-hand side cover. Using an ohmmeter or a multimeter set on the resistance scale, measure the resistance between the yellow leads (2-3 cylinders) and then the blue leads ((1-4 cylinders). In each case a resistance of 530 ± 50 Ohms at 20°C (68°F) should be indicated. Open or short circuits will be indicated by infinite or zero resistance respectively. A defective pulser must be renewed – there is no satisfactory means of repair.

5 Automatic timing unit (ATU): examination and renovation

1 The automatic timing unit rarely gives rise to problems unless of considerable age or if neglect has allowed the pivots to become corroded. As CDI ignition is fitted attention to this area is likely to be far from frequent. Although not specified as a routine servicing task, it is worth checking the unit annually to reduce the chance of failure. Access to the unit is gained by removing the circular inspection cover from the right-hand end of the crankcase, together with the ignition pickup assembly. Before loosening the pick-up backplate scribe a line from the plate to an adjacent portion of the housing casing and so mark the relative position of the backplate. This will allow reassembly without the need for ignition timing checking.

2 The unit is composed of two spring-loaded balance weights, which move outwards against spring tension as centrifugal force increases. At idling speed the force cannot overcome spring tension. Above idling speed the force is sufficient to throw the balance weights to the limit of their travel, ie full advance.

3 The balance weights must move freely on their pivots and be rust-free. The tension springs must also be in good condition. Keep the pivots lubricated and make sure the balance weights move easily, without binding. Most problems arise as a result of condensation, within the engine, which causes the unit to rust and balance weight movement to be restricted.

4   The automatic timing unit mechanism is fixed in relation to the crankshaft by means of a dowel pin. In consequence the mechanism cannot be replaced in anything other than the correct position. This ensures accuracy of ignition timing to within close limits.

5   The correct functioning of the auto-advance unit can be checked when the engine is running by the use of a stroboscopic light. If a strobe light is available, connect it to the ignition circuit of the 1 and 4 cylinders as directed by the manufacturer of the light. With the engine running, direct the beam of light at the fixed timing mark on the crankcase, through the aperture in the base plate. At tickover the 1.4 F-1 mark on the auto-advance unit should be precisely aligned with the fixed timing mark. When the engine is running at 2725 rpm (full advance) or above, the fixed timing mark should align between the two parallel lines which are marked on the auto-advance unit, slightly in advance of the 'F' mark. The above test relies, of course, on the static ignition timing being correct (Refer to Section 7 for the ignition timing procedure).

6   If the unit proves to be faulty it must be renewed if cleaning and lubrication fail to effect a repair. No replacement parts are available to overhaul the unit.

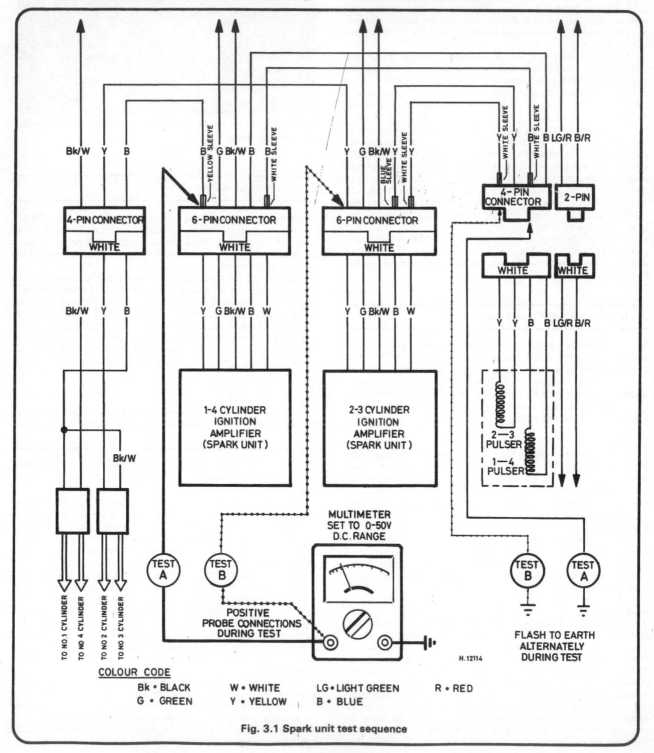

Fig. 3.1 Spark unit test sequence

5.3 Check balance weights, springs and pivots of ATU

6 Ignition coils: location and testing

1 Two separate ignition coils are fitted, each of which supplies a different pair of cylinders. The coils are mounted below the petrol tank, each side of the frame top tube.
2 If a weak spark, poor starting or misfiring causes the performance of the coils to be suspected, they should be tested by a Honda Service Agent or an auto-electrician who will have the appropriate test equipment.
3 It is unlikely that the coils will fail simultaneously. If intermittent firing occurs on one pair of cylinders the coils may be swopped over by interchanging the low tension terminal leads and the HT leads. If the fault then moves from one pair of cylinders to the other, it can be taken that the coil is faulty.
4 The coils are a sealed unit and therefore if a failure occurs, repair is impracticable. The faulty item should be replaced by a new component.

6.1 Each coil supplies a different pair of cylinders

7 Ignition timing: checking

1 The ignition timing is set accurately at the manufacturing stage and in the normal course of events will maintain its accuracy for an indefinite period. In the event that the timing setting has been lost in the course of overhaul, it can be checked either statically or dynamically, as described below.
2 The timing is best checked dynamically, using a stroboscope timing lamp. Start by removing the pulser inspection cover. Following the maker's instructions, connect the timing lamp to No 1 cylinder's high tension lead. Start the engine and allow it to idle. If the lamp is directed at the timing aperture in the pulser base plate, it will be noted that it appears to 'freeze' the timing marks on the rotating ATU. At the normal idle speed of 1050 ± 100 rpm, the 1.4 F-1 mark should be aligned with the fixed index mark.
3 If adjustment is required, slacken the pulser base plate screws. There are three of these screws around the periphery of the base plate. The base plate can now be rotated to set the timing position. Tighten the screws and re-check the timing as described above.
4 If the stroboscopic timing lamp is not available, the timing may be checked conveniently, if less accurately, as a static setting. Using the large hexagon on the crankshaft end, rotate the latter until the '1.4 S-F' mark lines up with the index mark. If the reluctor is viewed end-on, the projecting lug should align with a similar projection at the centre of the appropriate pickup or pulser coil. If necessary, adjust the pulser baseplate until the two coincide.

8 Sparking plugs: checking, cleaning and resetting the gaps

1 The Honda CB650 models are fitted with NGK or Nippon Denso (ND) sparking plugs of varying types. These are described in the Specification Section at the beginning of the Chapter. Note that in certain operating conditions, a change of plug grade may be required, though the standard grade will prove adequate in most cases.
2 Check and renew the sparking plugs at the intervals given in Routine Maintenance. This is the recommended maintenance interval given by the manufacturer, but reducing it somewhat would not be harmful. To reset the gap, bend the outer electrode closer to or further away from the central electrode until the gap is within the range of 0.6 – 0.7 mm (0.024 – 0.028 in) as measured with a feeler gauge. The gap is usually set to the lower figure to allow for erosion of the electrodes. Never bend the centre electrode or the insulator will crack, causing engine damage if the particles fall into the cylinder whilst the engine is running.
3 With some experience, the condition of the sparking plug electrodes and insulator can be used as a reliable guide to engine operating conditions.

4 Clean the electrodes with a wire brush and wipe the porcelain insulator. Check that the insulator is not broken or cracked, especially around the centre electrode. Check that the plug threads are clean, and the sealing washer is in position.
5 Always carry at least one spare sparking plug of the recommended grade. In the rare event of plug failure, this will allow immediate replacement and prevent inconvenience to the operator and strain on the engine when attempting to continue on one inoperative cylinder.
6 Beware of over-tightening the spark plugs, otherwise there is risk of stripping the threads from the aluminium alloy cylinder heads. The plugs should be sufficiently tight to seat firmly on their sealing washers, and no more. Use a spanner which is a good fit to prevent the spanner from slipping and breaking the insulator.
7 If the threads in the cylinder head strip as a result of over tightening the sparking plugs, it is possible to reclaim the head by means of a Helicoil thread insert. This is a cheap and convenient method of replacing the threads; most motorcycle dealers operate a service of this nature at an economical price.
8 Make sure the plug insulating caps are a good fit and have their rubber seals. They should be kept clean to prevent leakage and tracking of the HT current. These caps contain the suppressors that eliminate both radio and TV interference.

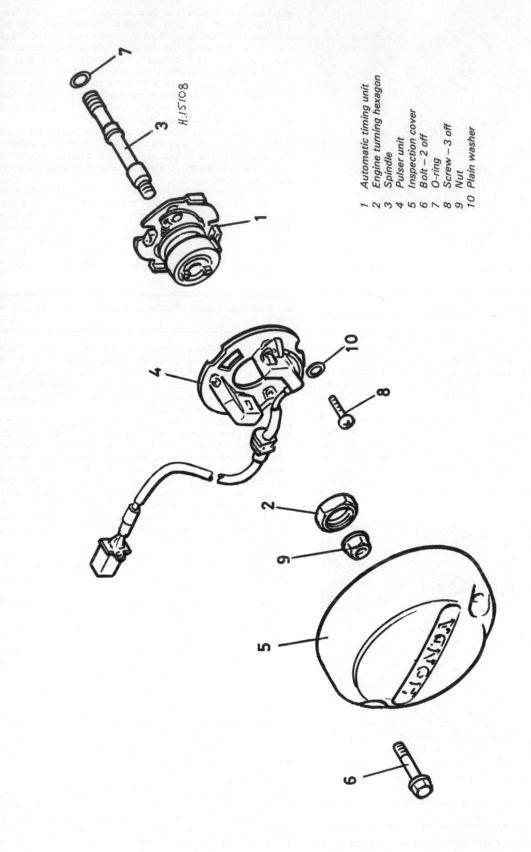

H.15108

1 Automatic timing unit
2 Engine turning hexagon
3 Spindle
4 Pulser unit
5 Inspection cover
6 Bolt – 2 off
7 O-ring
8 Screw – 3 off
9 Nut
10 Plain washer

Fig. 3.2 Ignition pulser assembly

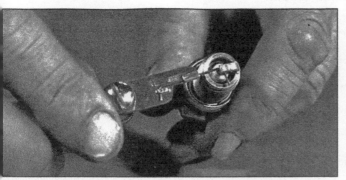

Electrode gap check - use a wire type gauge for best results

Electrode gap adjustment - bend the side electrode using the correct tool

Normal condition - A brown, tan or grey firing end indicates that the engine is in good condition and that the plug type is correct

Ash deposits - Light brown deposits encrusted on the electrodes and insulator, leading to misfire and hesitation. Caused by excessive amounts of oil in the combustion chamber or poor quality fuel/oil

Carbon fouling - Dry, black sooty deposits leading to misfire and weak spark. Caused by an over-rich fuel/air mixture, faulty choke operation or blocked air filter

Oil fouling - Wet oily deposits leading to misfire and weak spark. Caused by oil leakage past piston rings or valve guides (4-stroke engine), or excess lubricant (2-stroke engine)

Overheating - A blistered white insulator and glazed electrodes. Caused by ignition system fault, incorrect fuel, or cooling system fault

Worn plug - Worn electrodes will cause poor starting in damp or cold weather and will also waste fuel

9 Fault Diagnosis: ignition system

Symptom	Cause	Remedy
Engine will not start	Faulty ignition switch	Operate switch several times in case contacts are dirty. If lights and other electrics function, switch may need renewal
	Faulty CDI unit	Have unit tested. Replace if required
	Wiring fault	Check and repair wiring
	Starter motor not working	Discharged battery. Remove battery from machine and recharge
		Faulty starter circuit. Check for continuity
	Short circuit in wiring	Check whether fuse is intact. Eliminate fault before switching on again
	Completely discharged battery	If lights do not work, remove battery and recharge
Engine misfires	Fouled sparking plug	Renew plug and have original cleaned
	Power failure	Check pulsers, renew if required
Engine lacks power and overheats	Retarded ignition timing due to ATU unit failure	Check ATU
Engine 'fades' when under load	Pre-ignition	Check grades of plugs fitted: use recommended grades only

Chapter 4 Frame and forks

For information relating to 1981-on models refer to Chapter 7

Contents

Specifications

Frame

Type .... Tubular, double cradle

Front Forks

Type	Telescopic, hydraulically damped
Travel	142 mm (5.6 in)
Fork spring free length:	501.7 mm (19.8 in)
CB650 Z and A	501.7 mm (19.8 in)
CB650 CA	557.7 mm (21.96 in)
Service limit:	
CB650 Z and A	492.7 mm (19.4 in)
CB650 CA	541.0 mm (21.30 in)
Lower leg ID	35.042 – 35.104 mm (1.3796 – 1.3820 in)
Service limit	35.15 mm (1.384 in)
Stanchion OD	34.930 – 34.950 mm (1.3752 – 1.3760 in)
Service limit	34.90 mm (1.374 in)
Max stanchion bend	0.2 mm (0.008 in)
Fork oil capacity (per leg)	
Dry:	
CB650 Z	170 cc
CB650 A	175 cc
CB650 CA	209 cc
At oil change:	
CB650 Z	150 cc
CB650 A	155 cc
CB650 CA	190 cc
Fork oil grade	Automatic transmission fluid (ATF)

Rear suspension

Type	Swinging arm, welded tubular steel supported by oil filled suspension units
Travel:	
CB650 Z (UK), A and CA	91 mm (3.6 in)
CB650 Z (US)	77 mm (3.0 in)
Spring preload adjustment	5-position
Spring free length	224.7 mm (8.8 in)
Service limit	220.6 mm (8.7 in)

	CB650 Z	CB650 A and CA
Swinging arm bearing:		
Type	Plastic bush	Plastic bush
Swinging arm bush ID	21.500 – 21.552 mm	23.519 – 23.552 mm
	(0.8465 – 0.8485 in)	(0.9260 – 0.9272 in)
Service limit	21.7 mm	23.7 mm
	(0.854 in)	(0.93 in)
Inner sleeve OD	21.427 – 21.460 mm	23.427 – 23.460 mm
	(0.8436 – 0.8449 in)	(0.9223 – 0.9236 in)
Service limit	21.4 mm	23.4 mm
	(0.843 in)	(0.92 in)

Torque settings

Component	kgf m	lbf ft
Steering stem nut	8.0 – 12.0	58 – 87
Handlebar clamp	2.8 – 3.2	20 – 23
Fork clamp bolts (upper)	0.9 – 1.3	7 – 9
Fork clamp bolts (lower)	3.0 – 4.0	22 – 29
Fork cap bolt	2.0 – 3.0	15 – 22
Wheel spindle clamp bolts	1.8 – 2.5	13 – 18
Wheel spindle nut	5.5 – 6.5	40 – 47
Swinging arm pivot nut	6.0 – 7.0	43 – 51
Gearchange pedal pinch bolt	0.8 – 1.2	6 – 9
Brake drum torque arm nut	1.8 – 2.5	13 – 18
Rear suspension units (upper and lower)	3.0 – 4.0	22 – 29

1 General description

The Honda CB650 models employ a welded tubular steel frame of conventional full cradle design. This type of frame has the engine suspended inside the frame tubes rather than the engine composing any part of the frame. The right-hand lower cradle section can be removed to permit engine removal.

The front forks are hydraulically damped, consisting of two telescopic shock absorber assemblies, each of which comprises an inner tube, an outer tube, a spring and a cylinder, piston and valve. The whole fork assembly is attached to the frame by the steering head stem and is mounted on two bearing assemblies contained in the steering head housing.

The damping action of the fork is accomplished by the flow resistance of the fork oil flowing between the inner and outer tubes.

Rear suspension is by means of a pivoted rear fork or swinging arm, supported by a pair of oil damped rear suspension units. The rear suspension units incorporate 5-way adjustment of the spring preload to cope with varying riding requirements and conditions.

2 Front forks: method of removal

1 If necessary, it is possible to remove the forks as an assembly, together with the lower yoke. It is unlikely that this method will prove advantageous in view of the amount of preliminary dismantling necessary, but if this approach is deemed essential, follow the sequence detailed in Section 4, leaving the stanchions clamped in the lower yoke.

2 In most cases, the forks are best removed individually, without disturbing the yokes or steering head bearings. The relevant sequence is described in Section 3 of this Chapter.

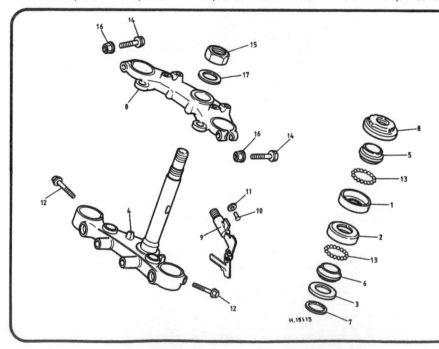

Fig. 4.1 Steering head assembly

1 Upper bearing cup
2 Lower bearing cup
3 Dust seal
4 Lower yoke/stem
5 Upper bearing cone
6 Lower bearing cone
7 Washer
8 Adjuster nut
9 Steering lock
10 Rivet
11 Washer
12 Pinch bolt – 2 off
13 Steel ball – 37 off
14 Pinch bolt – 2 off
15 Crown nut
16 Nut – 2 off
17 Washer

3 Front fork legs: removal and replacement

1   Place the machine securely on its centre stand, leaving adequate room around the front wheel area for comfortable working. It should be noted at this stage that some means of supporting the front wheel clear of the ground must be arranged. This can be accomplished by placing blocks or a jack beneath the sump, taking care not to damage the delicate sump fins. Alternatively a pair of tie-down straps, of the type used to secure motorcycles on trailers, can be arranged to lash the rear of the machine to the ground. Whatever method is chosen, make sure that there is no risk of the machine falling over whilst the front wheel and forks are removed.

2   On models equipped with twin front disc brakes, it will be necessary to release one of the calipers to permit front wheel removal. Note that if both legs are to be removed, both calipers should be released. This applies to single disc models where the left-hand fork leg is to be removed, even though wheel removal is possible with the caliper in place.

3   Slacken the caliper retaining bolts and lift the caliper clear of the forks. Insert a thin wooden wedge between the disc pads to prevent their displacement should the front brake lever be inadvertently operated prior to the calipers being refitted. The caliper(s) should be tied to the frame to avoid strain being placed on the hydraulic hose(s).

4   Release the speedometer cable by unscrewing the crosshead screw which secures it to the drive gearbox. Slacken the wheel spindle clamp nuts to release the wheel, which can now be lowered and disengaged from the front forks. Remove the four bolts which secure the front mudguard stays to the fork lower legs. Lift the mudguard clear and disengage the speedometer cable from its guide loop.

5   The fork legs can now be removed after releasing the upper and lower pinch bolts. Note that if it is intended to dismantle the forks, the fork top bolt should be slackened prior to removal. This is not easy with the forks in their normal position, as the handlebar obstructs the bolts. The easiest method is to slacken the pinch bolts, slide the stanchion down by about 3 inches then tighten the lower yoke pinch bolt. Prise off the plastic cap, and slacken the top bolt using the appropriate hexagon socket key. The pinch bolt can now be released and the fork leg lowered clear of the oke. Repeat the operation with the remaining fork leg, where required. If the yoke fork clamps prove to be excessively tight, they may be gently sprung, using a large screwdriver. This must be done with great care, to prevent breakage of the clamps, necessitating renewal of the complete yokes. Alternatively, if the fork legs appear to be binding in their clamps, such as may be the case if the machine has been involved in an accident and the fork legs are bent or twisted, liberal use of washing-up liquid applied to the clamps can be very beneficial.

6   If the fork legs are now to be dismantled further, for examination or renovation purposes, refer to Section 7 of this Chapter.

7   The fork legs can be refitted by reversing the above sequence, noting the following points. The fork top bolts should be tightened to 2.0 – 3.0 kgf m (15 – 22 lbf ft). This presents something of a problem unless a hexagon key socket is available, but can be overcome by sawing a short section off an ordinary key, and using a torque wrench and socket to drive it.

8   Fit each leg so that the edge of the plain face of each stanchion is flush with the top yoke. Tighten the lower yoke pinch bolts to 3.0 – 4.0 kgf m (22 – 29 lbf ft) followed by the upper yoke pinch bolts to 0.9 – 1.3 kgf m (7 – 9 lbf ft). When fitting the front wheel, note that the wheel spindle clamps are marked with an arrow and a letter F, denoting front. The front clamp nut should be tightened to the torque figure given below, followed by the rear nut, leaving a small gap to the rear of the spindle. The correct torque figure is 1.8 – 2.5 kgf m (13 – 18 lbf ft).

3.4a Release the speedometer cable securing screw

3.4b Two bolts in each leg retain the front mudguard

3.5a Release upper yoke pinch bolts and ...

3.5b ... lower yoke pinch bolts

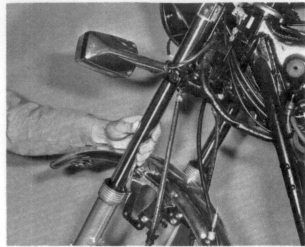

3.5c Slacken the fork top bolts prior to lowering the fork legs

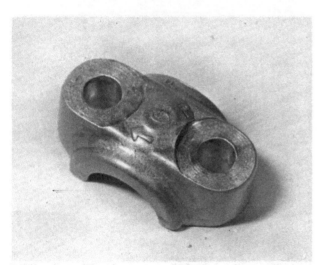

3.8 Note letter 'F' and arrow on spindle clamp; fit accordingly

4 Fork yokes and steering head bearings: removal and replacement

1   Fork yoke dismantling is an unwieldy process which is best avoided if at all possible. A considerable amount of work is involved in removing the numerous appendages from the fork yokes. For this reason, fork removal is best carried out as described in Section 3. It should be noted that some operations, such as inspection of the steering head bearings, can be undertaken without resorting to the full strip-down sequence given here, the trick being to disturb only those components which directly affect the part-dismantling sequences required. Before disconnecting any electrical connections the battery must be isolated by removing the lead from the positive terminal.
2   Start by removing the front wheel and fork legs as described in Section 3 of this Chapter. Although it is possible to leave the fork legs attached to the lower yoke, this makes an awkward job all the more difficult; they are best removed. The handlebar assembly must be removed, either completely after releasing the various switch harnesses, hydraulic pipe and cables, or by removing the handlebar clamp and displacing the assembly to provide clearance. In each case it is best to remove the fuel tank to avoid damage to the paintwork.

3   On all models, the handlebar is secured by a wide clamp which doubles as a fuse holder. Remove the fuse cover to expose the four socket screws. These should be released and the fuse holder displaced to release the handlebar. Move the handlebar assembly rearward to clear the yoke lodging it so that the master cylinder reservoir is not likely to leak.
4   If it is wished to remove the entire handlebar assembly, it will be necessary to release the handlebar switches, control cables and levers and the brake master cylinder. This is unlikely to be necessary in normal circumstances, except where accident damage has necessitated a full stripdown and examination of the frame, steering and suspension components.
5   Slacken and remove the screws which secure the headlamp assembly to the headlamp shell. Lift the unit away, disconnecting the bulb connector(s). The shell houses a number of multi-pin electrical connectors. These enter through a number of holes in the rear of the shell and will require separation before the shell can be released. Unscrew the reflectors to expose the headlamp shell retaining bolts, which can then be removed to permit headlamp shell removal.
6   Remove the instrument assembly by releasing the rubber-mounted retaining nuts and detaching the speedometer and tachometer drive cables. Lift the assembly clear together with the wiring harness and connector block. The headlamp/instrument mounting bracket can now be removed.
7   Work can now start on dismantling the steering head and yoke assembly. Note that some careful manoeuvring will be required to free the upper and lower yokes if the handlebar assembly has been displaced but not fully dismantled. Remove the steering stem cap nut, noting that Honda supply a special 30 – 32 mm double-ended socket, part number 07716–0020400, if required.
8   The upper yoke can now be removed, tapping it upwards to free it from the steering stem. With the upper yoke removed, the slotted adjustment nut will be exposed, and can be slacked using a C-spanner. As the nut is removed, the lower yoke and steering stem will drop downwards, and this should be supported until the nut is clear. With the nut removed, carefully lower the yoke and steering stem, catching the steel balls from the lower race as they drop free. A rag wrapped around the bearing area prior to the yoke removal is a useful aid to restraining the steel balls from their usual tendency to drop free, never to be seen again. The bearings in the upper race will usually remain in position. Reassembly is a reversal of the above sequence, noting that the steering head bearings should be checked, greased and adjusted as described in Sections 5 and 6.

5 Steering head bearings: examination and renovation

1   Before commencing reassembly of the forks, examine the steering head races. They are unlikely to wear out under normal circumstances until a high mileage has been covered. If, however, the steering head bearings have been maladjusted, wear will be accelerated. If, before dismantling, the forks had had a pronounced tendency to stick in one position when turned, often in the straight ahead position, the cups and cones are probably indented and need to be renewed.

2   Examine the cups and cones carefully; it is not necessary to remove the cups for this. The bearing tracks should be polished and free from indentations, cracks or pitting. If signs of wear are evident, the cups and cones must be renewed. In order for the straight line steering on any motorcycle to be consistently good, the steering head bearings must be absolutely perfect. Even the smallest amount of wear on the cups and cones may cause steering wobble at high speeds and judder during heavy front wheel braking. The cups and cones are an interference fit on their respective seatings and can be tapped from position using either the Honda special tool (No. 07953–3330000) or, if this is unavailable, a suitable long drift

3   Examine the ball bearings. Ball bearings are relatively cheap. If the originals are marked or discoloured they must be renewed as a complete set. To hold the steel balls in place during reassembly of the fork yokes, pack the bearings with grease. The upper race is fitted with eighteen $\frac{1}{4}$ inch (No 8) steel balls and the lower race, nineteen balls of the same size. Although each race has room for an extra steel ball it must not be fitted. The small gap allows the bearings to work correctly, stopping them skidding on each other and accelerating the rate of wear.

6 Steering head bearings: reassembly and adjustment

1   The steering head assembly should be reassembled in the reverse order of that given for dismantling in Section 4 from paragraph 7 onwards. When assembling the steering head, pack each bearing with grease prior to installation. Check that the correct number of balls is fitted to each bearing race. Ensure care is taken not to displace the balls during reassembly. The lower yoke is offered up and the upper bearing and adjuster nut is fitted finger tight. Adjustment of the bearings requires a C-spanner to fit the adjuster nut.

2   When fitting the steering stem adjuster nut, it must be adjusted so that it is a snug fit against the top cone, ie all perceptible play is taken up, and should then be backed off $\frac{1}{8}$ of a turn. Do not overtighten; it is easy to damage the head races by overtightening.

3   When correctly adjusted, it should be possible to move the steering from lock to lock with the lightest pressure on the handlebar end, with the front wheel clear of the ground. Final adjustment should be made after reassembly is completed.

7 Front fork legs: dismantling, renovation and reassembly

1   Having removed the fork legs as described in Section 3, they may be dismantled for further examination. Remember to slacken the fork top bolts **before** withdrawing the fork legs. If the correct sized Allen key is not available for releasing these top bolts, a bolt with a hexagon head of the required size, inserted into a socket spanner, will make an acceptable substitute. Always deal with one leg at a time and on no account interchange components from one leg to the other as the various moving parts will have bedded in during use, and should remain matched. Commence by draining the oil, either by way of the drain plug in the lower leg or by removing the top bolt and inverting the leg. Pumping the unit will assist in the draining operation.

2   Refer to the accompanying line drawing then start dismant-

ling, laying each part out on a clean surface as it is removed. Remove the chromium-plated top bolt and withdraw the fork spring. Using a 6 mm hexagon key, release the damper retaining bolt from the underside of the lower fork leg. It is helpful to clamp the lower leg in a vice during this operation, but care should be taken not to damage the soft alloy. Use soft jaws or wrap some rag around the leg to protect it.

3   Once the bolt has been released, the stanchion can be withdrawn from the lower leg. The alloy damper seat may remain in the lower leg, and should be shaken free. The damper assembly will remain inside the stanchion and can be tipped out of the upper end.

4   The front forks do not contain bushes. The fork legs slide directly against the hard chrome surface of the fork tubes. It follows, therefore, that the parts most liable to wear over an extended period of service are the internal surfaces of the lower leg and the outer surfaces of the fork stanchion or tube. If there is excessive play between these two parts they must be replaced as a complete unit. Honda recommend renewal of these components if either or both have reached the service limits given below.

Lower leg ID	35.15 mm (1.384 in)
Fork tube OD	34.90 (1.374 in)

Check the fork tube for scoring over the length which enters the oil seal. Bad scoring here will damage the oil seal and lead to fluid leakage.

5   It is advisable to renew the oil seals when the forks are dismantled even if they appear to be in good condition. This will save a strip-down of the forks at a later date if oil leakage occurs. The oil seal in the top of each lower leg is retained by an internal spring clip which can be prised out of position with a small screwdriver. A similar technique, using a larger screwdriver, can be employed to prise out the oil seal itself (see accompanying illustration). Exercise caution when removing oil seals; ensure the inner surface of the lower leg is not scratched or scored, and the top outer edge is not damaged when the screwdriver is bearing on this edge. Note that if the oil seals are not going to be renewed, they should not be disturbed. Damage will almost certainly be inflicted when prising the seals from position. If, prior to dismantling, fork oil had been seeping past either oil seal, then renewal of the seal is the only course of action. When installing new oil seals ensure that they are fitted correctly, with the closed side uppermost, and that they are dipped in clean fork oil prior to insertion in the fork lower leg. Drive the seals in squarely using a suitable drift, ie a large socket spanner, and secure them with the spring clip ensuring that the latter fits correctly in its groove.

6   Check that the dust excluder rubbers are not split or worn where they bear on the fork tube. A worn excluder will allow the ingress of dust and water which will damage the oil seal and eventually cause wear of the fork tube.

7   It is not generally possible to straighten forks which have been badly damaged in an accident, particularly when the correct jigs are not available. It is always best to err on the side of safety and fit new ones, especially since there is no easy means to detect whether the forks have been over stressed or metal fatigued. Any significant bend in the fork stanchions (tubes) will be detected by eye. If there is any doubt about straightness, the stanchions can be checked, after removal from the lower legs, by rolling them on a dead flat surface. Any misalignment will be immediately obvious.

8   The fork springs will take a permanent set after considerable usage and will need renewal if the fork action becomes spongy. The service limit for the total free length of each spring is given in the Specifications at the beginning of the Chapter. Always renew them as a matched pair.

9   Fork damping is governed by the viscosity of the oil in the fork legs, normally ATF. The piston ring fitted to the damper rod may wear if oil changes at the specified intervals are neglected. If damping has become weakened and does not improve as a

result of an oil change, the piston ring should be renewed. Check also that the oilways in the damper rod have not become obstructed.

10 When reassembling the fork legs, note the following points. The fork main spring should be fitted with tapered end downwards. Ensure the short rebound spring is fitted. Apply a little sealing compound to the threads of the socket head bolt at the fork leg bottom. Also note the copper sealing washer on this bolt and ensure it is refitted. Reinstall the drain plugs (if removed) and their copper sealing washers, prior to refilling the fork legs. Use the specified type and quantity of oil; do not overfill. Fit the fork springs with the closer wound coils uppermost. Do not omit the fork spring top seat. When fitting the fork top bolts, ensure the O-ring on the underside of the bolt head is in good condition.

12 To refit the fork legs to the yokes refer to Section 3, paragraphs 7 and 8 of this Chapter.

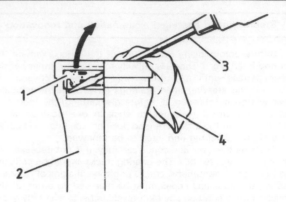

Fig. 4.2 Fork leg oil seal removal

| 1 | Oil seal | 3 | Lever |
| 2 | Leg | 4 | Protective padding |

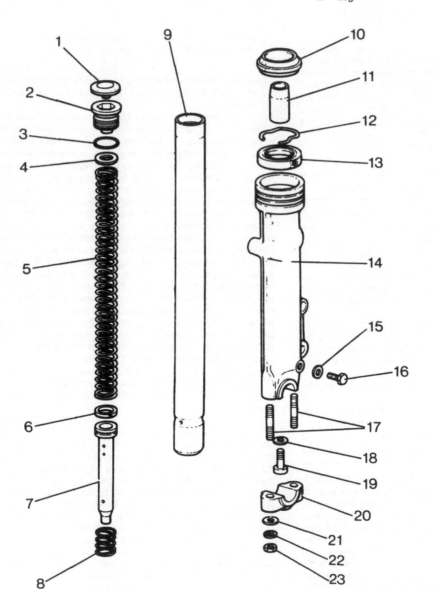

Fig. 4.3 Front forks

1 Cap
2 Top bolt
3 O-ring
4 Spring seat
5 Fork spring
6 Piston ring
7 Damper rod
8 Rebound spring
9 Stanchion
10 Dust excluder
11 Damper rod seat
12 Spring clip
13 Oil seal
14 Fork leg
15 Sealing washer
16 Drain plug
17 Stud – 2 off
18 Sealing washer
19 Allen bolt
20 Spindle clamp
21 Washer – 2 off
22 Spring washer – 2 off
23 Nut – 2 off

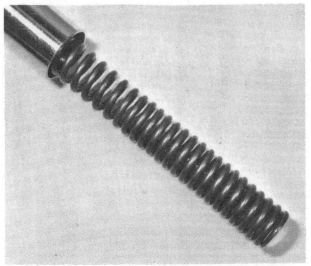

7.2a With the top bolt removed, withdraw the fork spring

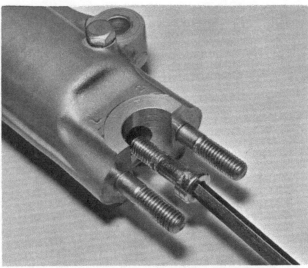

7.2b Unscrew and remove the socket bolt from the bottom of the lower leg

7.3 Withdraw stanchion from the lower leg

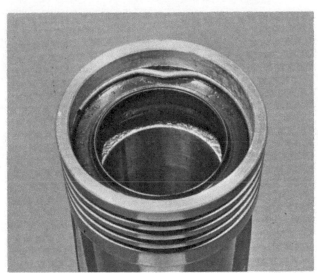

7.5 This spring clip retains the fork oil seal

7.6 Check the condition of the rubber dust excluders

7.10a Examine damper rod, spring and piston ring for wear

7.10b Piston ring should be renewed if damping action poor

7.11a Do not forget rebound spring when refitting damper rod

7.11b Ensure damper rod seat is refitted

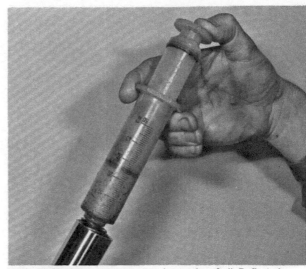

7.11c Refill with correct type and quantity of oil. Refit drain bolts first!

7.11d Fit spring top seat and ...

7.11e ... then refit fork top bolt; note O-ring

8 Steering head lock: maintenance

1   A security lock is mounted on the headstock, enabling the owner to immobilise the machine by locking the steering in one position. The lock consists of a key operated plunger which engages in a slot in the steering head. A small return spring disengages the lock mechanism when the key is released.
2   Maintenance is confined to keeping the lock lightly lubricated using light machine oil or one of the multipurpose aerosol lubricants. In the event that the lock malfunctions, it must be renewed; a repair is impracticable.

9 Frame: examination and renovation

1   The frame is unlikely to require attention unless accident damage has occurred. In some cases, renewal of the frame is the only satisfactory remedy if the frame is badly out of alignment. Only a few frame specialists have the jigs and mandrels necessary for resetting the frame to the required standards of accuracy, and even then there is no easy means of assessing to what extent the frame may have been over-stressed.
2   After the machine has covered a considerable mileage, it is advisable to examine the frame closely for signs of cracking or splitting at the welded joints. Rust corrosion can also cause weakness at these joints. Minor damage can be repaired by welding or brazing, depending on the extent and nature of the damage.

3   Remember that a frame which is out of alignment will cause handling problems and may even promote speed wobbles. If misalignment is suspected, as a result of an accident, it will be necessary to strip the machine completely so that the frame can be checked, and if necessary, renewed.

8.2 Keep steering lock lightly lubricated, especially in winter months

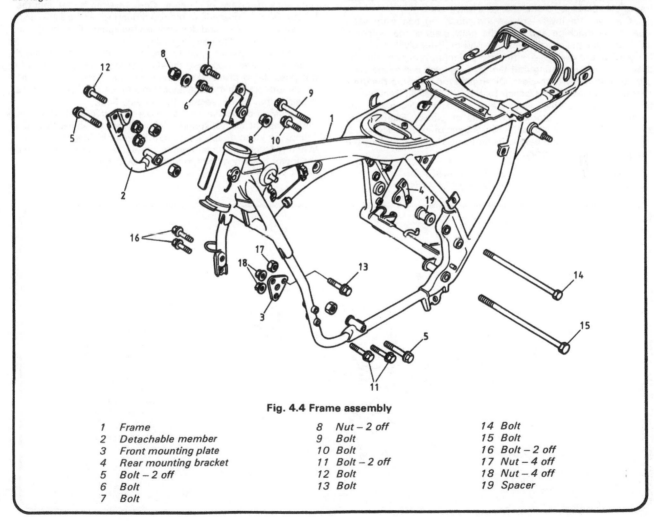

Fig. 4.4 Frame assembly

1	Frame	8	Nut – 2 off	14	Bolt
2	Detachable member	9	Bolt	15	Bolt
3	Front mounting plate	10	Bolt	16	Bolt – 2 off
4	Rear mounting bracket	11	Bolt – 2 off	17	Nut – 4 off
5	Bolt – 2 off	12	Bolt	18	Nut – 4 off
6	Bolt	13	Bolt	19	Spacer
7	Bolt				

10 Swinging arm fork: removal and renovation

1   The swinging arm fork is supported on two headed bushes which pivot on an inner sleeve. The assembly is retained by a long pivot shaft which passes through lugs on the frame and through the centre of the sleeve. A grease nipple is fitted to enable grease to be pumped to the bearing surfaces.

2   Wear in the swinging arm bushes is characterised by a tendency for the rear of the machine to twitch when ridden hard through a series of bends. This can be checked by placing the machine on the centre stand, and pushing the swinging arm from side to side. Any discernible free play will necessitate the removal of the swinging arm for further examination.

3   Commence removal operations, with the machine securely positioned on its centre stand, by removing the exhaust system, referring to Chapter 1, Section 4.11 for guidance. Detach the rear brake switch operating spring, then remove the brake adjusting nut. The brake operating rod can now be disengaged from the operating lever. Pull out the split pin from the torque arm mounting stud, then remove the securing nut and disengage the torque arm. Release the chainguard mounting nuts, and lift the guard away.

4   Remove the split pin which retains the rear wheel spindle nut, then slacken the nut. Release and back off the chain tensioner drawbolts, and swing them through 90° degrees. The wheel can now be pushed forwards as far as possible, and the final drive chain disengaged from the rear wheel sprocket. Slide out the stops from the fork ends to allow the wheel assembly to be withdrawn rearwards. Manoeuvre the wheel clear of the frame and place it to one side to await reassembly.

5   Remove the lower suspension mounting bolt from each side of the machine and push the units clear of the swinging arm. Slacken the pivot shaft nut and pull the pivot shaft out supporting the swinging arm. The swinging arm can now be drawn rearwards, noting that the caps on each end of the pivot tube will probably drop clear. Disengage the fork from the drive chain, and place it on a bench to await further dismantling.

6   Displace the pivot sleeve and wash the sleeve and the headed bushes to remove all traces of grease. Examine the sleeve and bush bearing surfaces for signs of wear or scoring. If damaged or worn, or if below the limits given below, replace the components as necessary.

    Bearing ID   21.7 mm (0.854 in)
    Sleeve OD   21.4 mm (0.843 in)

The two headed bushes may be driven out using a suitable bar or drift. It should be noted that if the bearings are to be removed for examination purposes only, new bearings should be available, as the original items may well be damaged or destroyed during removal. It follows, therefore, that the bearings should not be removed unless they are known to be worn out.

7   When fitting new bushes, ensure that they are tapped squarely into the swinging arm bore, and that they seat securely. Clean off any corrosion on the inner sleeve if this is to be re-used.

8   Reassemble in the reverse order of the above, lubricating the bushes with grease prior to installation. When assembly is complete, grease the pivot area by introducing grease through the grease nipple.

11 Rear suspension units: examination and dismantling

1   The Honda CB650 models are fitted with hydraulically-damped rear suspension units featuring 5-way spring preload adjustment. Adjustment is effected by turning an annular cam near the bottom of the units. Little in the way of maintenance is possible, the oil-filled damper units being sealed. It is possible to remove the springs if a suitable spring compressor, Honda part No 07959329001 or equivalent, is available.

2   Assemble the spring compressor, checking that it is fitted securely with no risk of slippage. Compress the spring until the locknut on the underside of the top mounting can be slackened. Unscrew the top lug and decompress the spring. The spring can now be lifted off the damper body.

3   Check the damper unit for signs of leakage, noting that if traces of oil are discovered the damper must be renewed. It is not possible to dismantle or overhaul the units. Measure the free length of each of the suspension springs, renewing them as a pair if they have reached the service limit.

4   Reassemble in the reverse order of that described above, noting that the closest spring coils must face downwards. When fitting the locknut, apply a trace of thread locking compound. When refitting the units, tighten the upper mounting nuts and the lower mounting bolts to a torque setting of 3.0 – 4.0 kgf m (22 – 29 lbf ft).

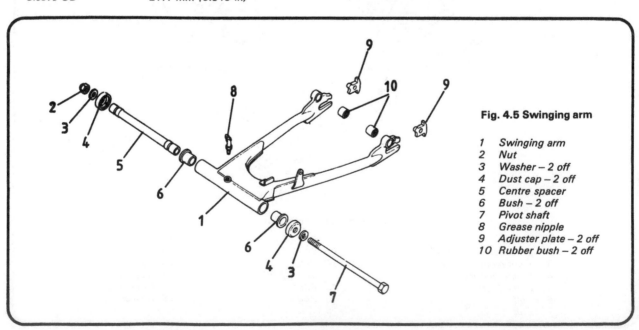

Fig. 4.5 Swinging arm

1   Swinging arm
2   Nut
3   Washer – 2 off
4   Dust cap – 2 off
5   Centre spacer
6   Bush – 2 off
7   Pivot shaft
8   Grease nipple
9   Adjuster plate – 2 off
10   Rubber bush – 2 off

10.3a Remove the forward bolt and ...

10.3b ... rear bolt to free the chainguard

10.5a Detach rear suspension units by removing lower retaining bolts

10.5b Withdraw the pivot shaft and ...

10.5c ... disengage the swinging arm

10.6a Dust caps will usually drop free as swinging arm is removed

10.6b Withdraw sleeve for examination

11.1 5-way adjustment of spring preload can be made here

10.6c Swinging arm fork: general view

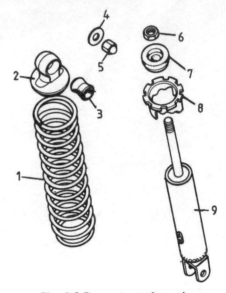
Fig. 4.6 Rear suspension unit

1	Spring	6	Nut
2	Mounting eye	7	Damping block
3	Rubber bush	8	Adjusting ring
4	Washer	9	Damper unit
5	Domed nut		

10.8 Lubricate pivot area with grease pumped in through grease nipple

12 Centre stand: examination

1   The centre stand pivots on a support tube which is clamped beneath the frame. A split pin is fitted at the clamped end of the tube as an additional security measure. One or two return springs retract the stand when not in use. The assembly should be checked for wear or damage, and the pivot tube greased. It is important that the stand remains secure as a failure would allow quite a lot of damage to be incurred if the machine should fall over. More importantly, the effects of a stand becoming displaced whilst the machine is being ridden would be catastrophic. This applies to the return springs, which should be renewed promptly if showing any signs of abrasion or serious corrosion.

13 Prop stand: examination

1   The prop stand is secured to a plate on the frame with a bolt and nut, and is retracted by either one or two extension springs. Make sure the bolt is tight and that each spring is not overstretched, otherwise an accident can occur if the stand drops during cornering. Check the rubber pad on the lower end of the side stand for excessive wear. The pad should be renewed when worn down to the wear limit line (below the arrow) at any point. When installing a new pad, ensure it is marked 'over 260 lbs only'.

14 Footrests: examination and renovation

1   The footrests are of the swivel type and are retained by a clevis pin secured by a split pin. The advantage of this type of footrest is that if the machine should fall over the footrest will fold up instead of bending.
2   If the footrests are damaged in an accident, it is possible to dismantle the assembly into its component parts by detaching each footrest from the rod and separating the folding rubber from the main support on which it pivots by withdrawing the split pin and releasing the clevis pin through the pivot. It is preferable to renew the damaged parts, but if necessary they can be bent straight by clamping them in a vice and heating to a dull red with a blow lamp whilst the appropriate pressure is applied. Never attempt to bend the footrests straight whilst they are still attached to the frame.
3   If heat is applied to the main footrest support during any straightening operation, it follows that the footrest rubber must be removed temporarily to prevent damage from heat.

15 Rear brake pedal: examination and renovation

1   The rear brake pedal is attached to a splined pivot shaft, and is secured by a single pinch bolt. A short arm welded to the inner end of the pivot shaft forms the point of attachment for the rod which operates the rear brake. The pivot is fitted with a strong return spring coiled around its boss, to provide the brake pedal with positive action.
2   If the brake pedal is bent or twisted, it can be drawn off the splined pivot and straightened by adopting the same technique as recommended for bent footrests. Note the position of the

15.1 Rear view of brake pedal on mounting plate (UK 650Z); note brake rod attachment and return spring

15.2 Note punch marks on brake pedal and shaft end

pedal in relation to the pivot splines before removal, so that it is replaced in the same position.
3   Check the return spring periodically and renew it if it shows signs of weakening.

16 Dualseat: removal and replacement

1   On CB650 Z models the seat is attached to a hinge plate by two pivot pins which fit into the seat lock assembly on one side and a hinge on the other side. The retaining assembly is secured to the base of the seat on the left-hand side. The pivot pins are affixed to the hinge plate. To remove the seat, release the spring loaded catch/seat lock and prop up the seat. Remove the bolt from the hinge plate and disengage the pivot pins. The seat mountings and damper rubbers can be left in place as the seat is lifted off.
2   The CB650 A and CA models share a simpler method of seat retention. An angled seat mounting bracket is fitted to the underside of the dualseat. The mounting bracket is secured to the left and right-hand of the machine by a single bolt.
3   If the dualseat is removed because it is torn, it is possible in most cases to find a specialist firm that recovers dualseats for an economical price, usually considerably cheaper than having to buy a new replacement. The usual charge is about 50% the cost of a new replacement, depending on the extent of the damage.

17 Instrument heads: removal and replacement

1   Unscrew the knurled retaining rings and disconnect the speedometer and tachometer drive cables. Remove the two domed nuts and rubber washers from the base of each instrument. Lift each head (one at a time!) and disconnect the leads to the bulb holders and then remove the instruments together.
2   The speedometer and tachometer heads cannot be repaired by the private owner, and if a defect occurs a new instrument has to be fitted. Remember that a speedometer in correct working order is required by law on a machine in the UK and also many other countries.
3   Before an instrument head is condemned as being faulty, check that the drive cable is not broken or kinked. In the latter case, the jerkiness produced will result in wildly inaccurate meter readings, even though the instrument itself may be undamaged.

16.1 Remove bolt and detach dualseat

17.1a Both instrument drive cables are secured by knurled rings

17.1b Remove domed nuts and rubber washers to separate instrument from its base

18 Instrument drive cables: examination and maintenance

1   It is advisable to detach the drive cables from time to time in order to check whether they are lubricated adequately, and whether the outer coverings are damaged or compressed at any point along their run. Jerky or sluggish movements can often be traced to a damaged drive cable.

2   For greasing, withdraw the inner cable. After removing all the old grease, clean with a petrol-soaked rag and examine the cable for broken strands or other damage.

3   Regrease the cable with a high melting point grease, taking care not to grease the last six inches at the point where the cable enters the instrument head. If this precaution is not observed, grease will work into the head and immobilise the instrument movement.

4   If any instrument head stops working suspect a broken drive cable. Inspection will show whether the inner cable has broken. Speedometer and tachometer cables are only supplied as a complete assembly. Make sure the cables are routed correctly through the clamps provided on the top fork yoke, brake branch pipe, and frame.

19 Speedometer and tachometer drives: location and examination

1   The speedometer is driven from a gear inside the front wheel hub assembly. The gear is driven internally by a tongued washer (receiver). The receiver engages with two slots in the wheel hub, on the left-hand side. As the whole gearbox is prepacked with grease on assembly the lubricant should last the life of the machine, or until new parts are fitted. The spiral pinion that drives off the internal gear is retained in the speedometer gearbox casing by a grub screw, which should always be secured tightly.

2   The tachometer drive runs off the camshaft in the cambox and screws directly into the cylinder head cover between cylinder Nos 3 and 4. The cable is secured to the cylinder head cover by a single bolt.

20 Cleaning the machine

1   After removing all surface dirt with a rag or sponge which is washed frequently in clean water, the machine should be allowed to dry thoroughly. Application of car polish or wax to the cycle parts will give a good finish, particularly if the machine receives this attention at regular intervals.

2   The plated parts should require only a wipe with a damp rag, but if they are badly corroded, as may occur during the winter when the roads are salted, it is permissible to use one of the proprietary chrome cleaners. These often have an oil base which will help to prevent corrosion from recurring.

3   If the engine parts are particularly oily, use a cleaning compound such as 'Gunk' or 'Jizer'. Apply the compound whilst the parts are dry and work it in with a brush so that it has an opportunity to penetrate and soak into the film of oil and grease. Finish off by washing down liberally, taking care that water does not enter the carburettors, air cleaners or the electrics.

4   If possible, the machine should be wiped down immediately after it has been used in the wet, so that it is not garaged in damp conditions which will promote rusting. Make sure that the chain is wiped and lightly greased or coated sparingly with heavyweight oil. Do not use one of the proprietary aerosol spray lubricants because these may be harmful to the O-rings used in the construction of the chains.

5   Check that the control cables are kept well oiled (this will only take 5 minutes of your time each week with an oil can). There is also less chance of water getting into the cables, if they are well lubricated.

19.1a Grease the tongued washer before ...

19.1b ... engaging the speedometer drive gearbox

21 Fault diagnosis: frame and forks

Symptom	Cause	Remedy
Machine veers to left or right with hands off handlebars	Wheels out of alignment Forks twisted Frame bent	Check wheels and realign Strip and repair Strip and repair or renew
Machine tends to roll at low speeds	Steering head bearings not adjusted correctly or worn	Check adjustment and renew the bearings, if worn
Machine tends to wander	Worn swinging arm bearings	Check and renew bearings Check adjustment and renew
Forks judder when front brake is applied	Steering head bearings slack Forks worn on sliding surfaces	Dismantle, lubricate and adjust Dismantle and renew worn parts
Forks bottom	Short of oil	Replenish with correct viscosity oil
Fork action stiff	Fork legs out of alignment Bent shafts, or twisted yokes	Strip and renew, or slacken clamp bolts, front wheel spindle and top bolts. Pump forks several times, and tighten from bottom upwards
Machine tends to pitch badly	Defective rear suspension units, or ineffective fork damping	Check damping action Check the grade and quantity of oil in the front forks

Chapter 5 Wheels, brakes and tyres

For information on 1981-on models refer to Chapter 7

Contents

Specifications

	CB650 Z (UK)	CB650 (US)	CB650 A	CB650 CA
Wheels				
Type	Comstar	Reversed Comstar	Wire-spoked	Reversed Comstar
Brakes				
Front	Twin hydraulic disc	Single hydraulic disc	Single hydraulic disc	Single hydraulic disc
Rear	Single leading shoe drum	Single leading shoe drum	Single leading shoe drum	Single leading shoe drum
Tyres				
Front	3.25H19 (4PR)	3.50H19 (4PR)	3.50H19 (4PR)	3.50S19 (4PR)
Rear	3.75H18 (4PR)	4.50H17 (4PR)	4.50H17 (4PR)	130/90S16
Tyre pressures (cold)				
Front, under all conditions	28 psi (2.0 kg/cm²)	28 psi (2.0 kg/cm²)	28 psi (2.0 kg/cm²)	28 psi (2.0 kg/cm²)
Rear, up to 90 kg (200 lb) load	28 psi (2.0 kg/cm²)	32 psi (2.25 kg/cm²)	32 psi (2.25 kg/cm²)	32 psi (2.25 kg/cm²)
Rear, above 90 kg (200 lb) load	40 psi (2.8 kg/cm²)	40 psi (2.8 kg/cm²)	36 psi (2.50 kg/cm²)	32 psi (2.25 kg/cm²)

Torque settings

Component	kgf m	lbf ft
Wheel spindle clamp nuts	1.8 – 2.5	13 – 18
Wheel spindle nut (front)	5.5 – 6.5	40 – 47
Wheel spindle nut (rear)	8.0 – 10.0	58 – 72
Rear wheel sprocket	8.0 – 10.0	58 – 72
Rear brake torque arm	1.8 – 2.5	13 – 18
Brake disc	2.7 – 3.3	20 – 24
Caliper bracket	3.0 – 4.0	22 – 29

1 General description

The Honda CB650 models employ various wheels and brake combinations, according to the model, the market on which it is to be sold, and the year of manufacture. Wheels are either of the conventional wire-spoked type or one of two patterns of the Honda Comstar design.

All models in the range are fitted with a 19 inch diameter front wheel. Rear wheel diameters vary considerably between models. The US market CB650 Z and CB650 A models are fitted with 17 inch diameter rear wheels. The UK market CB650 Z uses an 18 inch diameter rear wheel and the CB650 CA a 16 inch diameter rear wheel.

Tyre sizes vary accordingly with the different size wheels. All models, except the UK market CB650 Z, use a 3.50 inch section front tyre. These are all 'H' rated (safe up to speeds of 130 mph) except the item on the CB650 CA which is 'S' rated (safe up to 113 mph). The UK market CB650 Z model, is fitted with a 3.50 inch section, 'H' rated front tyre. The rear tyres fitted to the US market CB650 Z and CB650 A models are of 4.50 inch section and 'H' rating. The UK market CB650 Z model is fitted with an 'H' rated 3.75 inch section rear tyre. The CB650 CA utilises a low profile 130/90 mm section rear tyre, carrying an 'S' rating.

The CB650 A model is alone in the range in utilising traditional wire-spoked wheels. These have chromed steel rims laced to an alloy hub by butted wire spokes. The Comstar type wheels as fitted to the remaining models, differ in that they are built up with pressed spoke sections riveted to the hub and rim.

The standard design of Comstar wheel is fitted to the UK market CB650 Z model, the rest of the range use a reversed, black painted, pattern of Comstar.

Front brakes consist of either single (US market models) or twin (UK market CB650 Z model) hydraulic discs. Rear braking is provided by a single leading shoe drum on all models.

2 Front wheel: examination and renovation: wire spoked type

1 Place the machine on its centre stand so that the front wheel is clear of the ground. Spin the wheel by hand and check the rim for alignment, noting that the service limit is 2.0 mm (0.08 in). Small irregularities can be corrected by tightening the spokes in the affected area. Any flats in the wheel rim will be evident at the same time. In this latter case it will be necessary to have the wheel rebuilt with a new rim. The machine should not be run with a deformed wheel since this will have a very adverse effect on handling.

2 Check for loose or broken spokes. Tapping the spokes is a good guide to the correct tension; a loose spoke will always produce a different sound and should be tightened by turning the nipple in an anti-clockwise direction. Always check for run out by spinning the wheel again. If the spokes have to be tightened by an excessive amount, it is advisable to remove the tyre and tube as detailed in Section 23 of this Chapter. This will enable the protruding ends of the spokes to be ground off, thus preventing them from chafing the inner tube and causing punctures.

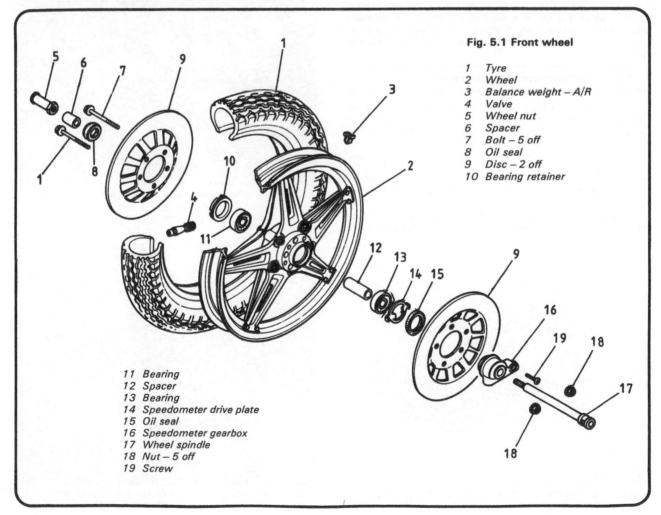

Fig. 5.1 Front wheel

1　Tyre
2　Wheel
3　Balance weight – A/R
4　Valve
5　Wheel nut
6　Spacer
7　Bolt – 5 off
8　Oil seal
9　Disc – 2 off
10　Bearing retainer

11　Bearing
12　Spacer
13　Bearing
14　Speedometer drive plate
15　Oil seal
16　Speedometer gearbox
17　Wheel spindle
18　Nut – 5 off
19　Screw

3 Front wheel: examination and renovation – Comstar type

1   The Comstar wheels, due to their design, must be considered as a single unit. Further to this, Honda do not offer any form of rebuilding facility. A number of private engineering shops offer this service, but it must be noted that Honda do not approve of this course of action.

2   Spin the wheel and check for rim alignment by placing a pointer close to the rim edge. If the total radial or axial alignment variation is greater than 2.00 mm (0.08 in) the manufacturers recommend that the wheel is renewed. This policy is, however, a counsel of perfection and in practice a larger runout may not affect the handling properties excessively. As remarked upon earlier, repair of a damaged wheel is not possible; the wheel must be renewed.

3   Check the rim for localised damage in the form of dents or cracks. The existence of even a small crack renders the wheel unfit for further use unless it is found that a permanent repair is possible using arc-welding. This method of repair is highly specialised and therefore the advice of a wheel repair specialist should be sought. Because tubeless tyres are used, dents may prevent complete sealing between the rim and tyre bead. This may not be immediately obvious until the tyre strikes a severely irregular surface, when the unsupported tyre wall may be deflected away from the rim, causing rapid deflation of the tyre. There again, specialist advice should be sought in order to establish whether continued use of the wheel is advisable.

4   Inspect the spoke blades for cracking and security. Check carefully the area immediately around the rivets which pass through the spokes and into the rim. In certain circumstances where steel spokes are fitted electrolytic corrosion may occur between the spokes, rivets and rim due to the use of different metals.

4 Front wheel: removal and replacement

1   With the front wheel supported well clear of the ground, remove the cross head screw which retains the speedometer cable to its drive gearbox, on the left-hand side of the hub. Pull the cable out and replace the screw, to prevent loss.

2   Remove the two bolts holding one of the caliper support brackets to the fork leg (twin disc models only) and lift the caliper and bracket assembly off the disc. Support the weight of the caliper with a length of string or wire attached to the frame or engine.

3   Slacken and remove the clamp nuts at the base of each fork leg. With the clamps released, the wheel will drop free and can be manoeuvred clear of the forks and mudguard.

4   Do not operate the front brake lever while the wheel is removed since fluid pressure may displace the pistons and cause leakage. Additionally, the distance between the pads will be reduced, making refitting of the brake discs more difficult.

5   Refit the wheel by reversing the dismantling procedure. Do not omit the spacer which is a push fit in the oil seal on the right-hand side of the wheel or the speedometer gearbox which is a push fit on the left-hand side. Ensure that the speedometer drive dogs engage with the notches in the gearbox drive sleeve. Lift the wheel into position and insert the wheel spindle so that the spindle head is flush with the outer face of the clamp and fork leg. Tighten the clamp front nut first to a torque setting of 1.8 – 2.5 kgf m (13 – 18 lbf ft). The rear nuts can now be tightened to the above setting, leaving a small gap at the **rear** of the clamp. Note that this is intentional, and thus the nuts must be secured in the above sequence. Refit the speedometer cable, and the brake caliper where appropriate. Secure the caliper retaining bolts to a torque setting of 3.0 – 4.0 kgf m (22 – 29 lbf ft).

6   Spin the wheel to ensure that it revolves freely and check the brake operation. Check that all nuts and bolts are fully tightened. If the clearance between the disc and pads is incorrect pump the front brake lever several times to adjust.

4.1 Separate speedometer cable at drive gearbox; refit screw to prevent loss

4.5a Refit spindle clamp with arrow mark and 'F' mark fitted accordingly

4.5b Tighten nuts and leave small gap at **rear** as shown

5 Front disc brake assembly: examination and brake pad renewal

1   Check the front brake master cylinder, hoses and caliper units for signs of leakage. Pay particular attention to the condition of the hoses, which should be renewed without question if there are signs of cracking, splitting or other exterior damage. Check the hydraulic fluid level by referring to the upper and lower level lines visible on the reservoir body.

2   Replenish the reservoir after removing the cap on the brake fluid reservoir and lifting out the diaphragm plate. The condition of the fluid can be checked at the same time. Checking the fluid level is one of the maintenance tasks which should **never be neglected**. If the fluid is below the lower level mark, brake fluid of the correct specification must be added. **Never** use engine oil or any fluid other than that recommended. Other fluids have unsatisfactory characteristics and will rapidly destroy the seals. The fluid level is unlikely to fall other than by a small amount as the disc pads gradually wear down, unless leakage has occurred somewhere in the system. If a rapid change of level is noted a careful check for leaks should be made before the machine is used again.

3   The two sets of brake pads should be inspected for wear. Each has a red groove, which marks the wear limit of the friction material. When this limit is reached, both pads in the set must be renewed, even if only one has reached the wear mark. A small inspection window, closed by a plastic cap, is provided in the top of each caliper unit so that examination of pad condition may be carried out easily.

4   If the brake action becomes spongy, or if any part of the hydraulic system is dismantled (such as when a hose has been renewed) it is necessary to bleed the system in order to remove all traces of air. Follow the procedure in Section 8 of this Chapter.

5   The brake pads can be removed after the inspection cover has been removed, this being retained by a single screw. Release the wire clip which retains the two pad holding pins, withdrawing the latter to free the pads. The pads and shim can be pulled upwards.

6   Refit the new pads and replace the caliper by reversing the dismantling procedure. The caliper piston should be pushed inwards slightly so that there is sufficient clearance between the brake pads to allow the caliper to fit over the disc. Do not omit the anti-chatter shim which should be fitted on the rear face of the piston side pad with the arrow pointing forward. It is recommended that both sides of the shim should be lightly coated with disc brake assembly grease (silicone grease). Use the grease sparingly and ensure that grease **does not** come into contact with the friction surface of the pad.

5.2 Remove cap and diaphragm to replenish reservoir

5.3 Line on pad denotes maximum wear limit

5.5a Remove screw and release inspection cover

5.5b Pull out the wire retaining clips and ...

5.5c ... withdraw the pad holding pins

5.5d The outer pad and ...

5.5e ... the inner pad and shim are now free to be lifted away

5.6 Refit, and lightly grease, shim as shown

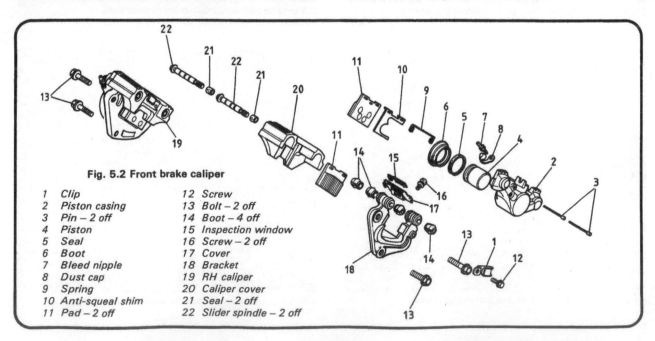

Fig. 5.2 Front brake caliper

1	Clip	12	Screw
2	Piston casing	13	Bolt – 2 off
3	Pin – 2 off	14	Boot – 4 off
4	Piston	15	Inspection window
5	Seal	16	Screw – 2 off
6	Boot	17	Cover
7	Bleed nipple	18	Bracket
8	Dust cap	19	RH caliper
9	Spring	20	Caliper cover
10	Anti-squeal shim	21	Seal – 2 off
11	Pad – 2 off	22	Slider spindle – 2 off

6 Front brake calipers: examination and renovation

1   Where twin disc brakes are fitted, it is advisable that each caliper is removed and overhauled separately, to prevent the accidental transposition of identical components. Select a suitable receptacle into which may be drained the hydraulic fluid. Remove the banjo bolt holding the hydraulic hose at the caliper and allow the fluid to drain. Take great care not to allow hydraulic fluid to spill onto paintwork; it is very effective paint stripper. Hydraulic fluid will also damage rubber and plastic components.

2   Remove the caliper from the fork leg and displace the brake pads as described in the preceding Section. Withdraw the two slider spindles and rubber boots from the support bracket.

3   Prise out the piston boot, using a small screwdriver, taking care not to scratch the surface of the cylinder bore. The piston can be displaced most easily by applying an air jet to the hydraulic fluid feed orifice. Be prepared to catch the piston as it falls free. Displace the annular piston seal from the cylinder bore groove.

4   Clean the caliper components thoroughly in trichlorethylene or in hydraulic brake fluid. **CAUTION**: Never use petrol for cleaning hydraulic brake parts otherwise the rubber components will be damaged. Discard all the rubber components as a matter of course. The replacement cost is relatively small and does not warrant re-use of components vital to safety. Check the piston and caliper cylinder bore for scoring, rusting or pitting. Measure the piston outside diameter with a micrometer. The service limit is 42.765 mm (1.6837 in). Also measure the internal diameter of the caliper cylinder bore; the service limit here is 42.915 mm (1.6896 in). If any of the aforementioned defects are evident, or the service limits have been reached, it is unlikely that a good fluid seal can be maintained and for this reason the components should be renewed. Inspect the slider spindles for wear and check their fit in the support bracket. Slack between the spindles and bores may cause brake judder if wear is severe.

5   To assemble the caliper, reverse the removal procedure. When assembling pay attention to the following points. Apply caliper grease (high heat resistance) to the rubber seals on the caliper spindles. If this grease is unavailable, use brake fluid. Do not damage the soft rubber of these seals when fitting them; they are surprisingly obstinate to fit. Also install new rubber boots in the support bracket with care. Apply a generous amount of brake fluid to the inner surface of the cylinder and to the periphery of the piston, then reassemble. Do not reassemble the piston with it inclined or twisted. When installing the piston push it slowly into the cylinder while taking care not to damage the piston seal. Apply brake pad grease to both sides of the anti-chatter shim. Bleed the brake after refilling the reservoir with new hydraulic fluid, then check for leakage while applying the brake lever tightly. Repeat the entire procedure for the second brake caliper. After a test run, check the pads and brake disc.

6   Note that any work on the hydraulic system must be undertaken in ultra-clean conditions. Particles of dirt will score the working parts and cause early failure.

7 Front disc brake master cylinder: examination and renovation

1   The master cylinder and hydraulic reservoir take the form of a combined unit mounted on the right-hand side of the handlebars, to which the front brake lever is attached. The master cylinder is actuated by the front brake lever, and applies hydraulic pressure through the system to operate the front brake when the handlebar lever is manipulated. The master cylinder pressurises the hydraulic fluid in the brake pipe which being incompressible, causes the piston to move in the caliper unit and apply the friction pads to the brake disc. If the master cylinder seals leak, hydraulic pressure will be lost and the braking action rendered much less effective.

2   Before the master cylinder can be removed, the system must be drained. Place a clean container below one caliper unit and attach a plastic tube from the bleed screw on top of the caliper unit to the container. Open the bleed screw one complete turn and drain the system by operating the brake lever until the master cylinder reservoir is empty. Close the bleed screw and remove the pipe. Care should be taken not to allow brake fluid to come into contact with paintwork; it is a very effective paintstripper. Brake fluid will also damage plastic and rubber components.

3   Remove the front brake stop lamp switch from the master cylinder (where fitted). Unscrew the union bolt and disconnect the connection between the brake hose and the master cylinder. Unscrew the two master cylinder fastening bolts and remove the master cylinder body from the handlebars. Empty any surplus fluid from the reservoir.

4   Remove the brake lever from the body, remove the boot stopper (taking care not to damage the boot) and then remove the boot. Remove the circlip that was hidden by the boot, the piston, primary cup, spring and check valve. Place the parts in a clean container and wash them in new brake fluid. Examine the cylinder bore and piston for scoring, scratches or nicks. Renew if any of these defects are apparent. Measure the master cylinder bore internal diameter and the piston outside diameter. They must be renewed if worn beyond the service limits given below.

> Master cylinder bore ID   14.055 mm (0.5533 in)
> Master cylinder piston OD   13.945 mm (0.5490 in)

Check also the brake lever for pivot wear, cracks or fractures, and the hose union threads and brake pipe threads for cracks or other signs of deterioration.

5   When reassembling the master cylinder follow the removal procedure in reverse order. Renew the various seals, lubricating them with hydraulic fluid before they are refitted. Make sure that the primary cup is fitted the correct way round. Mount the master cylinder on the handlebars so that the fluid reservoir is horizontal when the motorcycle is on the centre stand with the steering in the straight ahead direction. Fill with fresh fluid and bleed the system. Be sure to check the brake reservoir level. If the level is below the ring mark inside the reservoir, fill to the level with the prescribed brake fluid.

6   The component parts of the master cylinder assembly and the caliper assemblies may wear or deteriorate in function over a long period of use. It is however, generally difficult to foresee how long each component will work with proper efficiency. From a safety point of view it is best to change all the expendable parts every two years on a machine that has covered a normal mileage.

6.1 Remove banjo bolt from caliper. Take care not to spill any brake fluid

6.2 Withdraw the two caliper spindles

6.5a Fit new rubber seals here and ...

6.5b ... new rubber boots here, on the support bracket

6.5c Ensure all internal components are completely clean when reassembling

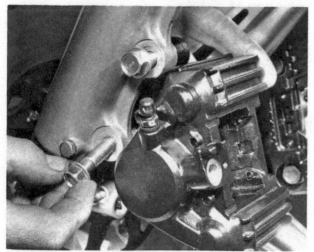

6.5d Refit caliper to fork leg and tighten mounting bolts

8   Bleeding the hydraulic brake system

1   If the hydraulic system has to be drained and refilled, if the front brake lever travel becomes excessive or if the lever operates with a soft or spongy feeling, the brakes must be bled to expel air from the system. The procedure for bleeding the hydraulic brake is best carried out by two persons.

2   First check the fluid level in the reservoir and top up with fresh fluid.

3   Keep the reservoir at least half full of fluid during the bleeding procedure.

4   Refit the diaphragm in the reservoir to prevent a spout of fluid or the entry of dust into the system. Place a clean glass jar below one caliper bleed screw and attach a clear plastic pipe from the caliper bleed screw to the container. Place some clean hydraulic fluid in the jar so that the pipe is always immersed below the surface of the fluid.

5   Unscrew the bleed screw one half turn and squeeze the brake lever. **Do not** release the lever until the bleeder valve is closed again. Repeat the operation a few times until no more air bubbles come from the plastic tube.

6   Keep topping up the reservoir with new fluid. When all the bubbles disappear, close the bleeder valve dust cap. Check the fluid level in the reservoir, after the bleeding operation has been completed. Repeat the air bleeding procedure with the second caliper unit (where appropriate).

7   Reinstall the diaphragm and refit the reservoir cap securely. Do not use the brake fluid drained from the system, since it will contain minute air bubbles.

8   Never use any fluid other than that recommended. Oil **must not be used** under any circumstances.

8.4 Clear plastic tube fitted to caliper bleed nipple for bleeding operation

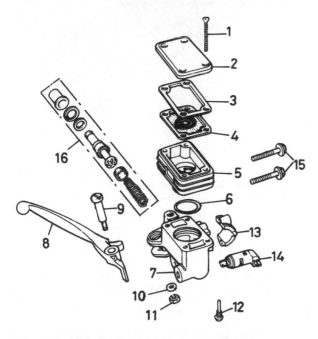

Fig. 5.3 Front brake master cylinder

1	Screw – 4 off	9	Pivot bolt
2	Cap	10	Washer
3	Seal	11	Nut
4	Diaphragm	12	Screw
5	Reservoir	13	Clamp
6	O-ring	14	Stop lamp switch
7	Master cylinder	15	Bolt – 2 off
8	Lever	16	Piston/seal set

9   Removing and replacing the brake discs

1   It is unlikely that the brake discs will require attention unless bad scoring has developed or warpage has occurred.

2   Warpage should be measured with the discs still attached to the wheel and the wheel in situ on the machine using a dial gauge. The maximum permissible warpage for the disc(s) is 0.3 mm (0.012 inch).

3   Disc removal is straightforward after the wheel has been taken out from the machine as described in Section 4 of this Chapter. Each disc is retained on the wheel hub by five bolts, which pass through the hub material. After removal of the bolts, each disc may be lifted off the hub boss.

4   The brake discs will wear eventually to a thickness which no longer provides sufficient support, and will probably begin to warp. The correct wear limit, which can be measured with a micrometer, is 6.0 mm (0.24 in), for single disc models, and 4.0 mm (0.16 in) for twin disc models.

5   When refitting the brake discs, tighten the five retaining bolts to a torque setting of 2.7 – 3.3 kgf m (20 – 24 lbf ft).

10 Front wheel bearings: examination and replacement

1   Access is available to the front wheel bearings when the speedometer gearbox has been removed from the left-hand side of the wheel and the spacer has been removed from the centre of the oil seal on the right. The bearings are of the ball journal type and non-adjustable. There are two bearings and two oil seals, the two bearings are interposed by a distance collar in the centre of the hub.

2   The bearing arrangement and dismantling procedure is similar irrespective of whether spoked, Comstar or reversed Comstar wheels are fitted. Note that a circular threaded retainer is fitted to the right-hand side of the hub. This retainer must be removed before the bearings can be released. A suitable peg spanner should be fabricated unless the correct Honda tool, part No 07710-0010200, is available. Do not resort to using a punch to loosen the retainer; this will only result in damage. The retainer will be staked in position and will require firm, even pressure to release it. Note also that new bearings and seals should be fitted whenever the old items are removed, so check carefully for wear before dismantling commences.

3   The left-hand bearing should be drifted out first, from the right-hand side of the wheel. Use a long drift against the inner face of the inner race. It may be necessary to knock the collar to one side so that purchase can be made against the race. Work round in a circle to keep the bearing square in the housing. The oil seal and speedometer drive dog plate will be pushed out as the bearing is displaced. After removal of the bearing, take out the distance collar and then drift out the right-hand bearing and oil seal in a similar manner, from inside the hub.

4   Remove all the old grease from the hub and bearings, wash the bearings in petrol, and dry them thoroughly. Check the bearings for roughness by spinning them whilst holding the inner track with one hand and rotating the outer track with the other. If there is the slightest sign of roughness renew them.

5   Before driving the bearings back into the hub, pack the hub with new grease and also grease the bearings. Use a tubular drift to drive the bearings back into position. Do not omit to refit the oil seals and distance collar.

6   Fit the retainer into position, having obtained a new replacement if there were signs of wear or damage to the threads. Tighten the retainer firmly, then secure it in this position by staking at the junction of the retainer and the wheel hub. A less brutal alternative to staking is the use of a small amount of locking fluid on the retainer threads.

9.4 Minimum thickness information on brake disc (UK 650Z shown)

10.1 Remove spacer from right-hand side of hub

10.2a Threaded bearing retainer must be removed before bearings

10.2b Use suitable home-made tool, such as is shown here, to remove bearing retainer

10.5 Note correct fitting of the bearings. Renew if at all suspect

11 Rear wheel: examination, removal and renovation

1   Place the machine on the centre stand so that the rear wheel is raised clear of the ground. Check for rim alignment, damage to the rim and loose or broken spokes by following the procedure relating to the front wheel, as described in Section 2 or 3 of this Chapter, depending on the wheel type fitted.

2   Detach the torque arm from the brake back plate after removing the nut secured by a split pin. Unscrew the brake rod adjuster nut fully and depress the brake pedal so that the rod leaves the trunnion in the brake operating arm. Refit the nut to secure the rod spring.

3   Remove the split pin from the end of the wheel spindle and slacken the wheel spindle nut. Release the chain adjuster locknuts, then slacken off the adjusters to allow them to be pushed down through 90°. Displace the small fillets from the fork ends.

4   Push the rear wheel forwards to allow the final drive chain to be disengaged from the rear wheel sprocket. If there is insufficient slack in the chain, as would be the case with a new chain, then it must be detached from the machine complete with the gearbox sprocket. Due to the one-piece construction of the chain ie, there is no split link, it cannot be separated. No

attempt should be made to 'split' the chain. The wheel can now be pulled rearwards to clear the swinging arm, and then manoeuvred free of the rear of the machine.

5   Refit the wheel by reversing the dismantling procedure. Ensure that the torque arm is secure and that the securing split pins are fitted. Likewise do not omit the wheel spindle nut securing pin. Adjust the rear brake as described in Section 14 of this Chapter, and check that the wheel alignment is correct.

11.3 Remove split pin and spindle nut and slacken chain adjuster locknuts

11.5 Ensure torque arm is secure and split pin installed

12 Rear wheel bearings: removal and replacement

1   With the rear wheel removed from the frame (refer to the previous Section), slacken the four driven sprocket retaining nuts, and displace the sprocket carrier by knocking it free of the cush drive rubbers in the hub. Note that both the sprocket carrier hub and the wheel hub feature a threaded retainer, similar to that fitted to the front hub (see Section 10). It follows that suitable peg spanner arrangements will be required unless the correct tools are available. The Honda part numbers for the latter are 07710-0010400 (retainer wrench body) 07710-0010100 (retainer wrench) and 07710-0010300 (re-

tainer wrench). The accompanying photographs show the methods used for retainer removal without the appropriate tools.

2   Unscrew the two retainers to allow the bearings to be driven out in a similar manner to that described in Section 10. Examination and reassembly details are also similar to those given for the front wheel components, noting the tabbed bearing central spacer, which locates the bearings, and the O-ring below the hub bearing retainer.

13 Rear drum brake assembly: examination and renovation

1   Access to the rear brake components is gained after removing the rear wheel as described in Section 11. With the wheel removed, the brake backplate assembly can be lifted clear of the drum.

2   Note that the brake operating lever is marked with punch marks as a guide to reassembly. Check the position of these prior to removal, if this proves necessary. Note that it is advisable to dismantle, clean and lubricate the brake cam each time the brake is overhauled.

3   Examine the brake linings for oil, dirt or grease. Surface dirt can be removed with a stiff brush but oil soaked linings should be renewed. High spots can be carefully eased down with emery cloth.

4   Examine the condition of the brake linings and if they have worn to less than 2.0 mm (0.08 in) in thickness they should be renewed. The brake linings are bonded to the brake shoes and thus separate linings are not available.

5   To remove the shoes displace the split pin from each fulcrum post and lift off the link plate. Remove the pinch bolt from the operating arm and pull it off the splined shaft followed by the wear indicator plate. Note the punch mark on the arm and shaft end to aid correct positioning on reassembly. Push the camshaft through the outside whilst simultaneously easing the brake shoe ends off the fulcrum posts at the opposite side of the brake back plate. After removal, displace the camshaft from between the shoe ends and separate the shoes from the springs.

6   Examine the drum surface for signs of scoring or oil contamination. Scoring occurs if the brake shoe linings have been allowed to get too thin. Both conditions will impair braking efficiency. Remove all traces of lining dust, taking care not to inhale any of it, as it is of an asbestos nature, and consequently hazardous to health. Use a clean rag soaked in petrol to remove all traces of grease and oil.

7   If deep scoring of the drum is evident, due to the linings having worn through to the shoe at some time, the drum must be skimmed on a lathe or renewed. Whilst there are firms who will undertake to skim a drum whilst fitted to the wheel, it should be borne in mind that excessive skimming will change the radius of the drum in relation to the brake shoes, therefore reducing the friction area until extensive bedding-in has taken place. Also full adjustment of the shoes may not be possible. If in doubt about this point, the advice of one of the specialist engineering firms who undertake this work should be sought.

8   Reassemble the brake back plate assembly by reversing the dismantling procedure. Before refitting existing shoes, roughen the lining surface sufficiently to break the glaze which will have formed in use. Check the return springs for wear or other damage at the hook ends and for stretching. Renew the springs, if necessary. Grease the operating camshaft and fulcrum posts with a high melting point grease before refitting the shoes. Note the seal on the camshaft, which prevents the escape of grease. When refitting the camshaft and arm, realign the marks to restore the original position. Refit the operating arm pinch bolt and tighten to a torque setting of 2.4 – 3.0 kgf m (17 – 22 lbf ft).

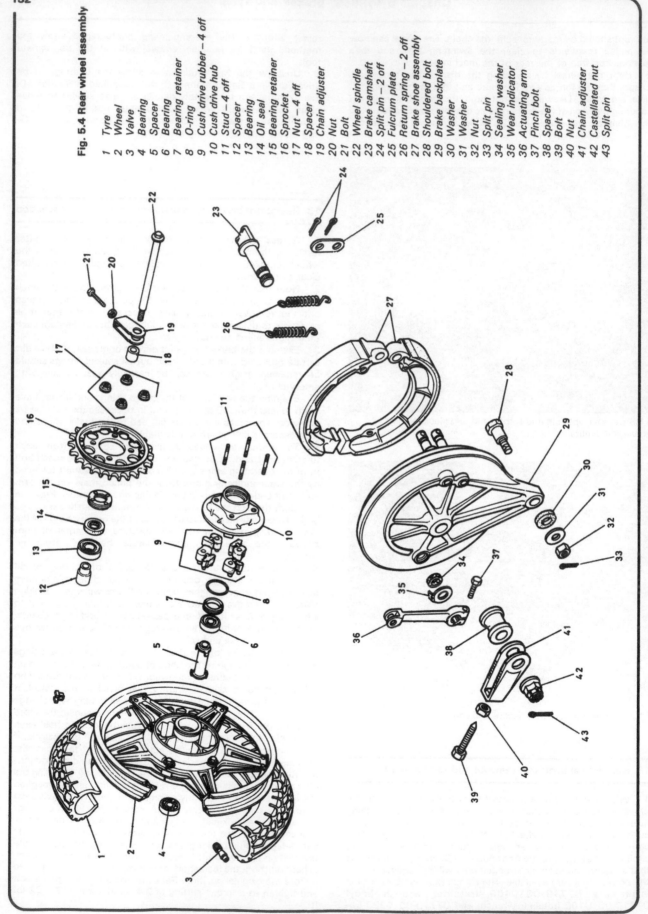

Fig. 5.4 Rear wheel assembly

1 Tyre
2 Wheel
3 Valve
4 Bearing
5 Spacer
6 Bearing
7 Bearing retainer
8 O-ring
9 Cush drive rubber – 4 off
10 Cush drive hub
11 Stud – 4 off
12 Spacer
13 Bearing
14 Oil seal
15 Bearing retainer
16 Sprocket
17 Nut – 4 off
18 Spacer
19 Chain adjuster
20 Nut
21 Bolt
22 Wheel spindle
23 Brake camshaft
24 Split pin – 2 off
25 Fulcrum plate
26 Return spring – 2 off
27 Brake shoe assembly
28 Shouldered bolt
29 Brake backplate
30 Washer
31 Washer
32 Nut
33 Split pin
34 Sealing washer
35 Wear indicator
36 Actuating arm
37 Pinch bolt
38 Spacer
39 Bolt
40 Nut
41 Chain adjuster
42 Castellated nut
43 Split pin

12.1a Separate sprocket and carrier from hub

12.1b Threaded retainers must be removed (note home-made tool) to ...

12.1c ... gain access to sprocket carrier hub and wheel hub bearings

12.2a Sprocket carrier bearings can be drifted out using a socket spanner

12.2b Fit new bearings if original items at all suspect

12.2c Note the flanged bearing central spacer and ...

12.2d ... the O-ring around the hub bearing housing

12.2e Refit the threaded retainers. Note oil seal function of retainers

12.2f Reinstall short spacer in sprocket carrier hub centre

12.2g Tighten four nuts down to correct torque setting

13.1 Lift away brake backplate assembly for examination

13.2 Note punch mark and wear indicator (arrowed)

Tyre changing sequence - tubeless tyres

Deflate tyre. After releasing beads, push tyre bead into well of rim at point opposite valve. Insert lever adjacent to valve and work bead over edge of rim.

Use two levers to work bead over edge of rim. Note use of rim protectors.

When first bead is clear, remove tyre as shown.

Before fitting, ensure that tyre is suitable for wheel. Take note of any sidewall markings such as direction of rotation arrows.

Work first bead over the rim flange.

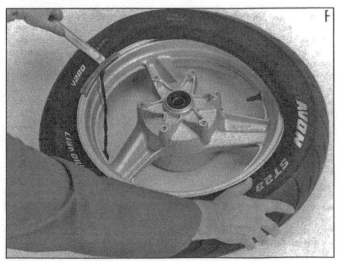

Use a tyre lever to work the second bead over rim flange.

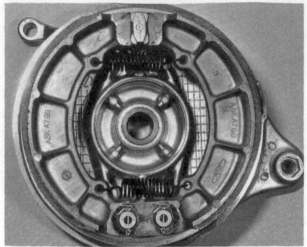

13.4 Brake linings are bonded to the shoes and cannot be separated

13.5 Remove split pins and link plate to remove brake shoes

13.7 Internal diameter of the drum must not exceed figure stated here

13.8 Grease operating cam when refitting shoes

14 Adjusting the rear brake

1   Adjustment of the rear brake is correct when there is 20 – 30 mm ($\frac{3}{4}$" – 1" approx) up and down movement measured at the rear brake pedal foot piece, between the fully off and on position. Adjustment is carried out by turning the nut on the brake rod.

2   The height of the brake pedal when at rest may be adjusted by means of the stop bolt which passes through a plate welded to the pedal shank. Loosen the locknut before making the adjustment and tighten it when adjustment is complete.

3   Either adjustment may require the brake pedal operated stop lamp switch to be re-adjusted.

15 Rear sprocket assembly: examination, renovation and replacement

1   The final driven sprocket is held to the rear wheel by four flanged nuts. The sprocket needs to be replaced only if the teeth are worn, hooked or chipped.

2   Although becoming an extremely expensive ideal, it is always a good policy to change both sprockets, and the chain, at the same time. If all three components are not replaced simultaneously, rapid wear will result on the new components.

16 Rear cush drive: examination and renovation

1   The cush drive assembly is contained in the left-hand side of the wheel hub. It takes the form of four, curiously shaped, rubber blocks. These engage with vanes on the coupling which is bolted to the rear sprocket. The rubbers engage with ribs on the hub and the whole assembly forms a shock absorber which permits the sprocket to move within certain limits. This cushions any surge or roughness in the transmission which would otherwise convey an impression of harshness.

2   The usual sign that shock absorber rubbers are worn is excessive movement in the sprocket, or rubber dust appearing in between the sprocket and hub. The rubbers should then be taken out and renewed.

14.2 Adjust brake pedal height with locknut and stop bolt

16.2 Renew cush drive rubbers as a set if worn or damaged

17 Final drive chain: examination and lubrication

1 As the final drive chain is fully exposed on all models it requires lubrication and adjustment at regular intervals. To adjust the chain, take out the split pin from the rear wheel spindle and slacken the spindle nut. Undo the torque arm bolt, and leave the bolt in position, slacken the chain adjuster, locknuts and turn the adjusters inwards to tighten the chain, or outwards to slacken the chain.

2 Chain tension is correct if there is 15 – 25 mm ($\frac{5}{8}$ – 1 inch) slack measured at the centre of the bottom run of the chain between the two sprockets. Note that excessive free play, 40 mm (1$\frac{5}{8}$ inch) or more, will allow the chain to contact the frame, causing damage to both. Marks on the chain adjusters must be in line with the vertical marks on the frame rear fork to align the wheel correctly. If there is any doubt as to the accuracy of these marks, a final check can be made by laying a straight wooden plank alongside the wheels, each side in turn. See the accompanying diagram.

3 On the later CB650 models (A and CA), as a guide to determining the condition of the chain whilst it is still in position on the machine, small coloured stickers are affixed to the rearmost part of the chain adjusters. If the first indicator mark (red) aligns with the end of the swinging arm, then the chain must be replaced, as it is beyond its useful life. The second mark (green) indicates the position of the swinging arm, with a new chain fitted, in relation to the chain adjusters.

4 Do not run the chain too tight to try to compensate for wear, or it will absorb a surprising amount of engine power. Also it can damage the gearbox and rear wheel bearings. As stated earlier, a chain that is run exceptionally slack is equally undesirable, causing damage to the frame, excessive wear on the sprockets and rendering the machine a danger to ride.

5 All models are equipped with endless chains, ie they have no split link. The earlier, CB650 Z (1979) models, were fitted with chains of traditional design relying on the operator for lubrication. The CB650 A and CA models, however, are equipped with 'O-ring type' final drive chains. These chains are lubricated upon assembly; the grease lubricant is then sealed in for the life of the components by the use of O-rings placed at each end of the rollers. These chains are also of a larger size than their traditional counterparts.

6 With all models utilising endless chains, it is not possible to periodically remove the chain for full lubrication in 'Linklyfe' or 'Chainguard' as with other chains. Even if periodical removal was a practicable operation, the molten lubricant should not be applied to the O-ring type chains, because of the damage the heat would cause to the O-rings. Lubrication on the CB650 Z models is, therefore, restricted to frequent cleaning and re-

lubrication with one of the good quality aerosol chain lubricants.

7 Although the internal bearing surfaces of the O-ring type chain are permanently lubricated, the outside of the rollers and the sprockets with which they mesh are not. Lubricant must, therefore, be applied at regular intervals to prevent wear of the sprockets and chain rollers. Use either a high-melting point grease applied sparingly, or a heavyweight gear oil, SAE 80 or 90, for correct lubrication. **Do not** use one of the proprietary aerosol chain lubricants (unless specifically manufactured for use on O-ring chains) as O-ring damage may occur. **Do not** use any solvents, other than paraffin, to clean the chain, as these too may damage the O-rings.

8 To check if the chain is due for renewal, lay it lengthwise in a straight line and compress it endwise until all play is taken up. Anchor one end, then pull in the opposite direction to take up the play which has developed. If the chain extends by more than $\frac{1}{4}$ inch per foot, it should be renewed.

9 Removal of the final drive chain for renewal, necessitates the removal of the rear wheel and the complete swinging arm assembly as described in Chapter 4, Section 10. Once the swinging arm has been removed, the chain may be lifted off the gearbox sprocket. When fitting a new chain, ensure that the gearbox and rear wheel sprockets are in good condition, if new sprockets are not being fitted.

11 A British-made replacement chain for the Honda CB650 is available from Renold Ltd.

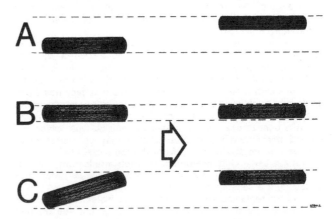

Fig. 5.5 Method of wheel alignment

B – Correct   A and C – Incorrect

17.2 Use vertical marks to ensure correct wheel alignment after chain adjustment

18 Tubed tyres: removal and replacement – wire spoked wheels

1   At some time or other the need will arise to remove and replace the tyres, either as a result of a puncture or because replacements are necessary to offset wear. To the inexperienced, tyre changing represents a formidable task, yet if a few simple rules are observed and the technique learned, the whole operation is surprisingly simple.

2   To remove the tyre from either wheel, first detach the wheel from the machine. Deflate the tyre by removing the valve core, and when the tyre is fully deflated, push the bead from the tyre away from the wheel rim on both sides so that the bead enters the centre well of the rim. Remove the locking ring and push the tyre valve into the tyre itself.

3   Insert a tyre lever close to the valve and lever the edge of the tyre over the outside of the rim. Very little force should be necessary: if resistance is encountered it is probably due to the fact that the tyre beads have not entered the well of the rim, all the way round.

4   Once the tyre has been edged over the wheel rim, it is easy to work round the wheel rim, so that the tyre is completely free from one side. At this stage the inner tube can be removed.

5   Now working from the other side of the wheel, ease the other edge of the tyre over the outside of the wheel rim that is furthest away. Continue to work around the rim until the tyre is completely free from the rim.

6   If a puncture has necessitated the removal of the tyre, reinflate the inner tube and immerse it in a bowl of water to trace the source of the leak. Mark the position of the leak, and deflate the tube. Dry the tube, and clean the area around the puncture with a petrol soaked rag. When the surface has dried, apply rubber solution and allow this to dry before removing the backing from the patch, and applying the patch to the surface.

7   It is best to use a patch of self vulcanizing type, which will form a permanent repair. Note that it may be necessary to remove a protective covering from the top surface of the patch after it has sealed into position. Inner tubes made from a special synthetic rubber may require a special type of patch and adhesive, if a satisfactory bond is to be achieved.

8   Before replacing the tyre, check the inside to make sure that the article that caused the puncture is not still trapped inside the tyre. Check the outside of the tyre, particularly the tread area to make sure nothing is trapped that may cause a further puncture.

9   If the inner tube has been patched on a number of past occasions, or if there is a tear or large hole, it is preferable to

discard it and fit a replacement. Sudden deflation may cause an accident, particularly if it occurs with the front wheel.

10   To replace the tyre, inflate the inner tube for it just to assume a circular shape but only to that amount, and then push the tube into the tyre so that it is enclosed completely. Lay the tyre on the wheel at an angle, and insert the valve through the rim tape and the hole in the wheel rim. Attach the locking ring on the first few threads, sufficient to hold the valve captive in its correct location.

11   Starting at the point furthest from the valve, push the tyre bead over the edge of the wheel rim until it is located in the central well. Continue to work around the tyre in this fashion until the whole of one side of the tyre is on the rim. It may be necessary to use a tyre lever during the final stages.

12   Make sure there is no pull on the tyre valve and again commencing with the area furthest from the valve, ease the other bead of the tyre over the edge of the rim. Finish with the area close to the valve, pushing the valve up into the tyre until the locking ring touches the rim. This will ensure that the inner tube is not trapped when the last section of bead is edged over the rim with a tyre lever.

13   Check that the inner tube is not trapped at any point. Reinflate the inner tube, and check that the tyre is seating correctly around the wheel rim. There should be a thin rib moulded around the wall of the tyre on both sides, which should be an equal distance from the wheel rim at all points. If the tyre is unevenly located on the rim, try bouncing the wheel when the tyre is at the recommended pressure. It is probable that one of the beads has not pulled clear of the centre well.

14   Always run the tyres at the recommended pressures and never under or over inflate. The correct pressures for solo use are given in the Specifications Section of this Chapter.

15   Tyre replacement is aided by dusting the side walls, particularly in the vicinity of the beads, with a liberal coating of french chalk. Washing up liquid can also be used to good effect, but this has the disadvantage of causing the inner surface of the wheel rim to rust.

16   Never replace the inner tube and tyre without the rim tape in position. If this precaution is overlooked there is a good chance of the ends of the spoke nipples chafing the inner tube and causing a crop of punctures.

17   Never fit a tyre that has a damaged tread or sidewalls. Apart from legal aspects, there is a very great risk of a blowout, which can have very serious consequences on a two-wheeled vehicle.

18   Tyre valves rarely give trouble, but it is always advisable to check whether the valve itself is leaking before removing the tyre. Do not forget to fit the dust cap, which forms an effective extra seal.

19 Valves and valve caps: tubed tyres

1   Inspect the valves in the inner tubes from time to time making sure that the seal and spring are making an effective seal. There are tyre valve tools available for clearing damaged threads in the valve body and incorporating thread clearing for the outside thread of the body. A key is also incorporated for tightening the valve core.

2   The valve caps prevent dirt and foreign matter from entering the valve, and also form an effective second seal so that in the event of the tyre valve sticking, air will not be lost.

3   Note that when a dust cap is fitted for the first time to a balanced wheel the wheel may have to be rebalanced.

20 Tubeless tyres: removal and replacement – Comstar wheels

1   It is strongly recommended that should a repair to a tubeless tyre be necessary, the wheel is removed from the machine and taken to a tyre fitting specialist who is willing to

do the job or taken to an official Honda dealer. This is because the force required to break the seal between the wheel rim and tyre bead is considerable and considered to be beyond the capabilities of an individual working with normal tyre removing tools. Any abortive attempt to break the rim to bead seal may also cause damage to the wheel rim, resulting in an expensive wheel replacement. If, however, a suitable bead releasing tool is available, and experience has already been gained in its use, tyre removal and refitting can be accomplished as follows.

2   To remove the tyre from either wheel, first detach the wheel from the machine by following the procedure in Sections 4 or 11 depending on whether the front or the rear wheel is involved. Deflate the tyre by removing the valve insert and when it is fully deflated, push the bead of the tyre away from the wheel rim on both sides so that the bead enters the centre well of the rim. As noted, this operation will almost certainly require the use of a bead releasing tool.

3   Insert a tyre lever close to the valve and lever the edge of the tyre over the outside of the wheel rim. Very little force should be necessary; if resistance is encountered it is probably due to the fact that the tyre beads have not entered the well of the wheel rim all the way round the tyre. Should the initial problem persist, lubrication of the tyre bead and the inside edge and lip of the rim will facilitate removal. Use a recommended lubricant, a dilute solution of washing-up liquid or french chalk. Lubrication is usually recommended as an aid to tyre fitting but its use is equally desirable during removal.

4   The risk of wheel rim damage from levers can be minimised by the use of proprietary plastic rim protectors placed over the rim flange at the point where the tyre levers are inserted. Suitable home-made rim protectors can be fabricated from short lengths (say 6 inch) of thick-walled plastic fuel pipe, split down one side. This approach may be adopted whenever levers are being used and, therefore, the risk of damage is likely.

5   Once the tyre has been edged over the wheel rim. It is easy to work around the wheel rim so that the tyre is completely free on one side.

6   Working from the other side of the wheel, ease the other edge of the tyre over the outside of the wheel rim which is furthest away. Continue to work around the rim until the tyre is freed completely from the rim.

7   Refer to the following Section for details relating to puncture repair and the renewal of tyres. See also the remarks relating to the tyre valves in Section 2.

8   Refitting of the tyre is virtually a reversal of the removal procedure. If the tyre has a balance mark (usually a spot of coloured paint), as on the tyres fitted as original equipment, this must be positioned alongside the valve. Similarly, any arrow indicating direction of rotation must face the right way.

9   Starting at the point furthest from the valve, push the tyre bead over the edge of the wheel rim until it is located in the central well. Continue to work around the tyre in this fashion until the whole of one side of the tyre is on the rim. It may be necessary to use a tyre lever during the final stages. Here again, the use of a lubricant will aid fitting. It is recommended strongly that when refitting the tyre only a recommended lubricant is used because such lubricants also have sealing properties. Do not be over generous in the application of lubricant or tyre creep may occur.

10 Fitting the upper bead is similar to fitting the lower bead. Start by pushing the bead over the rim and into the well at a point diametrically opposite the tyre valve. Continue working round the tyre, each side of the starting point, ensuring that the bead opposite the working area is always in the well. Apply lubricant as necessary. Avoid using tyre levers unless absolutely essential to help reduce damage to the soft wheel rim. The use of the levers should be required only when the final portion of bead is to be pushed over the rim.

11 Lubricate the tyre beads again prior to inflating the tyre, and check that the wheel is evenly positioned in relation to the tyre beads. Inflation of the tyre may well prove impossible without the use of a high pressure air hose. The tyre will retain air completely only when the beads are firmly against the rim

edges at all points and it may be found when using a foot pump that air escapes at the same rate as it is pumped in. This problem may also be encountered when using an air hose, on new tyres which have been compressed in storage and by virtue of their profile hold the beads away from the rim edges. To overcome this difficulty, a tourniquet may be placed around the circumference of the tyre, over the central area of the tread. The compression of the tread in this area will cause the beads to be pushed outwards in the desired direction. The type of tourniquet most widely used consists of a length of hose closed at both ends with a suitable clasp fitted to enable both ends to be connected. An ordinary tyre valve is fitted at one end of the tube so that after the hose has been secured around the tyre it may be inflated, giving a constricting effect. Another possible method of sealing the beads to obtain initial inflation is to press the tyre into the angle between a wall and the floor. With the air line attached to the valve, pressure is then applied on the tyre by the hand and shin, as shown in the accompanying illustration. This will often effect an initial seal between tyre and wheel allowing inflation to occur.

12 Having successfully accomplished inflation, increase the pressure to 40 psi and check that the tyre is evenly disposed on the wheel rim. This may be judged by checking that the thin positioning line found on each tyre wall is equidistant from the rim around the total circumference of the tyre. If this is not the case, deflate the tyre, apply additional lubrication and reinflate. Minor adjustments to the tyre position may be made by bouncing the wheel on the ground.

13 Always run the tyre at the recommended pressures and never under or over-inflate. The correct pressures for solo use are given in the Specification Section of this Chapter. If a pillion passenger is carried increase the rear tyre pressure only as recommended.

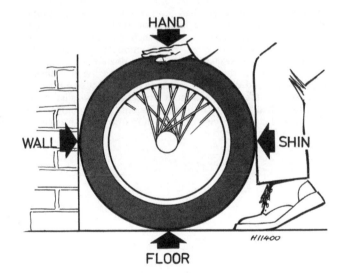

Fig. 5.6 Tubeless tyre initial inflation

21 Puncture repair and tyre renewal

1   The primary advantage of the tubeless tyre is its ability to accept penetration by sharp objects such as nails etc without loss of air. Even if loss of air is experienced, because there is no inner tube to rupture, in normal conditions a sudden blow-out is avoided.

2   If a puncture of the tyre occurs, the tyre should be removed

for inspection for damage before any attempt is made at remedial action. The temporary repair of a punctured tyre by inserting a plug from the outside should not be attempted. Although this type of temporary repair is used widely on cars, the manufacturers strongly recommend that no such repair is carried out on a motorcycle tyre. Not only does the tyre have a thinner carcass, which does not give sufficient support to the plug, but the consequences of a sudden deflation are often sufficiently serious that the risk of such an occurrence should be avoided at all costs.

3   The tyre should be inspected both inside and out for damage to the carcass. Unfortunately the inner lining of the tyre – which takes the place of the inner tube – may easily obscure any damage and some experience is required in making a correct assessment of the tyre condition.

4   There are two main types of tyre repair which are considered safe for adoption in repairing tubeless motorcycle tyres. It should be noted, however, that neither method is satisfactory for holes greater than 3 mm diameter. The first type of repair consists of inserting a mushroom-headed plug into the hole from **inside** of the tyre. The hole is prepared for insertion of the plug by reaming and the application of an adhesive. The second repair is carried out by buffing the inner lining in the damaged area and applying a cold or vulcanised patch. The patch should have a diameter of at least 25 mm and the hole should be treated with a rubber sealant to prevent the ingress of moisture which could cause damage to the carcass fabric. Because both inspection and repair, if they are to be carried out safely, require experience in this type of work, it is recommended that the tyre be placed in the hands of a repairer with the necessary skills, rather than repaired in the home workshop.

5   In the event of an emergency, the only recommended get-you-home repair is to fit a standard inner tube of the correct size. If this course of action is adopted, care should be taken to ensure that the cause of the puncture has been removed before the inner tube is fitted. It will be found that the valve hole in the rim is considerably larger than the diameter of the inner tube valve stem. To prevent the ingress of road dirt, and to help support the valve, a spacer should be fitted over the valve. A conversion spacer for most Honda models equipped with Comstar wheels is available from Honda dealers.

6   In the event of the unavailability of tubeless tyres, ordinary tubed tyres fitted with inner tubes of the correct size may be fitted. Refer to the manufacturer or a tyre fitting specialist to ensure that only a tyre and tube of equivalent type and suitability is fitted, and also to advise on the fitting of a valve nut/spacer to the rim hole.

22 Tyre valves: description and renewal – Comstar wheels

1   It will be appreciated from the preceding Sections that the adoption of tubeless tyres has made it necessary to modify the valve arrangement, as there is no longer an inner tube which can carry the valve core. The problem has been overcome by using a moulded rubber valve body which locates in the wheel rim hole. The valve body is pear-shaped, and has a groove around its widest point which engages with the rim forming an airtight seal.

2   The valve is fitted from the rim well, and it follows that it can only be renewed whilst the tyre itself is removed from the wheel. Once the valve has been fitted, it is almost impossible to remove it without damage, and so the simplest method is to cut it as close as possible to the rim well. The two halves of the old valve can then be removed.

3   The new valve is fitted by inserting the threaded end of the valve body through the rim hole, and pulling it through until the groove engages in the rim. In practice, a considerable amount of pressure is required to pull the valve into position, and most tyre fitters have a special tool which screws onto the valve end to enable purchase to be obtained. It is advantageous to apply a little tyre bead lubricant to the valve to ease its insertion. Check that the valve is seated evenly and securely.

4   The incidence of valve body failure is relatively small, and leakage only occurs when the rubber valve case ages and begins to perish. As a precautionary measure, it is advisable to fit a new valve when a new tyre is fitted. This will preclude any risk of the valve failing in service. When purchasing a new valve, it should be noted that a number of different types are available. The correct type for use in the Comstar wheel is a Schraeder 412, Bridgeport 193M, or equivalent.

5   The valve core is one of the same type as that used with tubed tyres, and screws into the valve body. The core can be removed with a small slotted tool which is normally incorporated in plunger type pressure gauges. Some valve dust caps incorporate a projection for removing valve cores. Although tubeless tyre valves seldom give trouble, it is possible for a leak to develop if a small particle of grit lodges on the sealing face. Occasionally, an elusive slow puncture can be traced to a leaking valve core, and this should be checked before a genuine puncture is suspected.

6   The valve dust caps are a significant part of the tyre valve assembly. Not only do they prevent the ingress of road dirt into the valve, but also act as a secondary seal which will reduce the risk of sudden deflation if a valve core should fail.

23 Fault diagnosis: wheels, brakes and tyres

Symptom	Cause	Remedy
Handlebars oscillate at low speed	Buckle or flat in wheel rim, most probably front wheel	Check rim for damage by spinning wheel Renew wheel if not true
	Tyre pressure incorrect	Check, and if necessary adjust
	Tyre not straight on rim	Check tyre. If necessary, deflate tyre and reposition
	Worn wheel or steering head bearings	Check and renew or adjust
Machine tends to weave	Tyre pressure incorrect	Check and if necessary adjust. If sudden, check for puncture
	Suspension worn or damaged	Check action of front forks and rear suspension units. Check swinging arm for wear.
Machine lacks power and accelerates poorly	Front disc brake binding	Hot disc or caliper indicates binding Overhaul caliper(s) and master cylinder, fit new pads if required, check disc(s) for scoring or warpage
	Rear drum brake binding	Hot drum indicates binding. Dismantle brake. Clean and lubricate pivots. Check springs. Check drum for scoring or warpage
Brakes grab or judder when applied gently	Rear drum brake: ends of linings not chamfered	File chamfer on leading edge of lining
	Rear disc brake: caliper bracket worn or loose; disc warped	Dismantle, clean and renew as required
	Brake drum scored or warped	Remove wheel. Skim drum or renew
	Front brake: pads badly worn or scored	Renew pads and check disc and caliper
	Wrong type of pad fitted Warped disc	Renew
Brakes squeal Front disc(s):	Glazed pads. Pads worn to backing metal	Sand pad surface to remove glaze, then use brake gently for about 100 miles to permit bedding in. If worn to backing check that disc is not damaged and renew as necesssary
	Caliper and pads polluted with brake dust or foreign matter	Dismantle and clean. Overhaul caliper where necessary
Rear drum:	Accumulated brake dust in drum	Dismantle and clean
	Glazed linings	Restore as described for glazed pads
Excessive front brake lever travel	Air in system	Find cause of air's presence. If due to leak, rectify, then bleed brake
	Very badly worn pads	Renew, and overhaul system where required
	Badly polluted caliper	Dismantle and clean
Front brake lever feels springy	Air in system	See above
	Pads glazed	See above
	Caliper jamming	Dismantle and overhaul

Chapter 6 Electrical system

For information relating to 1981-on models refer to Chapter 7

Contents

Specifications

Battery
Make	Yuasa
Capacity	12 volt 12 Ah
Polarity	Negative earth

Alternator
Type	Three-phase
Output	260W at 5000 rpm

Voltage regulator/rectifier
Type	Integrated circuit, non-adjustable

Starter motor
Brush length	12.0 – 13.0 mm (0.47 – 0.51 in)
Service limit	7.5 mm (0.3 in)

Bulbs
	CB650 Z (UK)	CB650 Z (US)	CB650 A and CA
Headlamp	60/55W H4 Quartz Halogen	65/50W QH Sealed beam unit	65/50W QH Sealed beam unit
Parking lamp	4W	N/A	N/A
Tail/stop lamp	5/21W x 2	8/27W x 1	8/27W x 2
Registration plate lamp	IOW	N/A	N/A
Directon indicator lamps	21W x 4	23W x 2 (rear)	23W x 2 (rear)
Running/indicator lamps	N/A	8/23W x 2 (front)	8/23W x 2 (front)
Indicator warning lamp	3.4W x 2	3.4W x 2	3.4W x 2
Oil pressure warning lamp	3.4W	3.4W	3.4W
Neutral indicator lamp	3.4W	3.4W	3.4W
High beam indicator lamp	3.4W	3.4W	3.4W
Instrument illumination lamps	3.4W x 4	3.4W x 4	3.4W x 4

All bulbs rated at 12V

Fuses
		CB650 Z (UK)	CB650 Z (US)	CB650 A and CA
Main		30A	30A	30A
1	(Headlamp)	10A	10A	10A
2	(Oil/neutral (UK) Oil/neutral, running (US))	10A	10A	10A
3	(Horn/stop/indicators)	10A	10A	10A
4	(Tail/instruments)	10A	10A	10A

1 General description

All models are equipped with a twelve volt electrical system powered by a crankshaft-mounted alternator (ac generator). An electromagnetic rotor is retained on the left-hand crankshaft end by means of a taper and central securing bolt. The stator and brush assemblies are retained inside the outer casing.

The resulting alternator current (ac) is fed to a combined regulator/rectifier unit mounted below the dualseat or behind the left-hand side panel depending on the model. This device converts the output to direct current (dc) and controls the system voltage. The regulated and rectified supply is then fed to the battery and electrical system.

2 Testing the electrical system: general

1   Checking the electrical output and the performance of the various components within the charging system requires the use of test equipment of the multi-meter type and also an ammeter of 0 – 5 ampere range. When carrying out checks, care must be taken to follow the procedures laid down and so prevent inadvertent incorrect connections or short circuits. Irreparable damage to individual components may result if reversal of current or shorting occurs. It is advised that unless some previous experience has been gained in auto-electrical testing the machine be returned to a Honda Service Agent or auto-electrician, who will be qualified to carry out the work and have the necessary test equipment.

2   If the performance of the charging system is suspect the system as a whole should be checked first, followed by testing of the individual components to isolate the fault. The three main components are the alternator, the rectifier, and the regulator. Before commencing the tests, ensure that the battery is fully charged, as described in Section 6.

3 Charging system: checking the output

1   A quick check of the charging system condition may be made using the above-mentioned ammeter and voltmeter arrangement. Raise the dualseat and remove the side panels to gain access to the electrical components. Disconnect the red battery positive (+) lead, attaching it to the positive (+) terminal of the ammeter. Run a lead from the negative (–) ammeter terminal to the positive (+) battery terminal. Set the multimeter to the 0 – 20 volts dc scale (or higher) and connect the positive (+) probe to the positive (+) battery terminal. The negative (-) probe should be earthed.

2   Start the engine and allow it to reach normal running temperature, then turn on the main lights, set on main beam. At 1650 rpm, the discharge reading found at idle speed should be cancelled out. If the engine speed is now increased to 5000 rpm, the ammeter should show a zero reading or a slight charge, with 14 volts indicated on the voltmeter or multi-meter.

3   If the output is erratic or noticeably below the specified amount, either the alternator or the rectifier may be at fault. The rectifier may be tested as described in Section 4.

4   Poor output may be caused by sticking of the alternator pick-up brushes or by worn brushes. If either brush is worn to below the scribed limit line both brushes should be renewed.

5   The alternator stator should be tested as follows, using a multi-meter set to the resistance function. Disconnect the block connector which connects the three alternator stator output leads plus the brush leads to the regulator/rectifier unit. The connector is located below the dualseat (above the battery) on CB650Z models and behind the left-hand side panel on CB650 A and CA models. Using the multi-meter check that continuity exists between all three wires when tested in pairs. Check also that no lead has continuity with earth. If the results of the check do not correspond with those specified, there is evidence of short-circuits or open-circuits in the stator windings

or the leads. The specified resistance is 0.41 – 0.51 ohms, though it will be sufficient just to note whether there is an open or short circuit.

4 Regulator/rectifier unit: location and testing

1   If performance of the charging system is suspect, but the alternator is found to be in good condition, it is probable that one side of the combined regulator/rectifier unit is malfunctioning. Exactly which side is malfunctioning is of academic interest only because the sealed unit cannot be repaired, but must be renewed.

2   Voltage regulator performance may be checked using a voltmeter connected across the battery. Connect the positive voltmeter lead to the positive (+) battery terminal and the negative voltmeter lead to the negative (–) battery terminal. Start the engine and increase the speed until a 14 volt output is registered. Increase the engine speed to approximately 5000 rpm. If the regulator is functioning correctly, the voltage will not rise to above 15 volts. If this voltage is exceeded, the unit is malfunctioning.

3   A further test of the unit's condition may be made by measuring the resistance of the rectifier circuit. To check these values accurately a meter or meters capable of reading in ohms and kilo ohms will be required. Failing this, most multimeters will be able to give a reliable indication despite reading solely on the latter scale. It is generally sufficient to distinguish between a very high resistance and a very low resistance when checking diode condition.

4   Before the test is started, it is helpful to understand what the rectifier is required to do. We have established that it converts ac to dc for charging purposes, and it does this by a matrix of diodes. Diodes, represented by a triangular symbol with a bar across one end in the accompanying diagram, can be imagined to act as one way valves, passing current in one direction only. Thus, in our diagram the current should flow in the direction indicated by the point of the diode symbol, but must not flow back past the bar. In the event of rectifier failure, one or more of the diodes will have failed to operate correctly and thus will either pass current in both directions or will set up a high resistance in both directions. The purpose of the test is to check that the diodes are functioning correctly.

5   Trace the leads from the finned regulator/rectifier unit situated below the dualseat or behind the left-hand side panel. These terminate in a 6-pin and a 4-pin connector block, each of which should be separated. Using an ohmmeter or a multimeter set on resistance, check the resistance between the green lead (4-pin connector) and each of the yellow leads in turn (6-pin connector). Reverse the meter probes and check the resistance between the green lead and each of the yellow leads in the opposite direction. If these diodes are in good condition there should be very little resistance (approx 5 – 40 ohms) in the normal direction of current flow and very high resistance (at least 2000 ohms) in the reverse direction.

6   Test the remaining diodes by connecting the meter between the red/white lead (4-pin connector) and each of the yellow leads (6-pin connector) in turn, noting the resistance reading obtained and then reversing the meter probes to obtain readings in the opposite direction. If these diodes are in good condition results should be the same as those given above. It follows that if in any of the tests the meter shows a very high or very low resistance in both directions then that diode is faulty and the complete regulator/rectifier unit must be renewed.

5 Battery: examination and maintenance

1   All models are fitted with a 12 volt battery of 12 ampere hour capacity.

2   On the CB650 Z models, the transparent plastic case of the battery permits the upper and lower levels of the electrolyte to be observed without disturbing the battery, with the right-hand

side panel removed. On the CB650 A and CA models, with the battery repositioned in the centre of the frame tubes, hidden from view, an upper and lower level mark is visible with the left-hand side panel removed. Maintenance is normally limited to keeping the electrolyte level between the prescribed upper and lower limits and making sure that the vent tube is not blocked. The lead plates and their separators are also visible through the transparent case, a further guide to the general condition of the battery.

3   Unless acid is spilt, as may occur if the machine falls over, the electrolyte should always be topped up with distilled water to restore the correct level. If acid is spilt onto any part of the machine, it should be neutralised with an alkali such as washing soda or baking powder and washed away with plenty of water, otherwise serious corrosion will occur. Top up with sulphuric acid of the correct specific gravity (1.260 to 1.280) only when spillage has occurred. Check that the vent pipe is well clear of the frame or any of the other cycle parts.

4   It is seldom practicable to repair a cracked battery case because the acid present in the joint will prevent the formation of an effective seal. It is always best to renew a cracked battery, especially in view of the corrosion which will be caused if the acid continues to leak.

5   If the machine is not used for a period, it is advisable to remove the battery and give it a refresher charge every six weeks or so from a battery charger. If the battery is permitted to discharge completely, the plates will sulphate and render the battery useless.

6   Occasionally, check the condition of the battery terminals to ensure that corrosion is not taking place and that the electrical connections are tight. If corrosion has occurred, it should be cleaned away by scraping with a knife and then using emery cloth to remove the final traces. Remake the electrical connections whilst the joint is still clean, then smear the assembly with petroleum jelly (NOT grease) to prevent recurrence of the corrosion. Badly corroded connections can have a high electrical resistance and may give the impression of a complete battery failure.

6   Battery: charging procedure

1   The normal charging rate for batteries should always be one tenth of the battery's rated capacity. The 12 amp hour battery will, therefore, require charging at 1.2 amps. It is permissible to charge at a slightly higher charging rate. Honda recommend a maximum charging rate of 1.4 amps, if necessary. A higher charging rate should be avoided as this will shorten the working life of the battery, and if excessive, may overheat the battery and buckle its plates.

2   Always remove the vent caps when recharging a battery, otherwise the gas created within the battery when charging takes place may burst the case with disastrous consequences.

7   Fuses: location and replacement

1   The Honda CB650 models are equipped with a total of five fuses. The main fuse is housed in an enclosed fuse holder on the side of the starter solenoid. It is of the ribbon type and is rated at 30 amps. A spare fuse element is carried in the fuse holder. The remaining fuses are housed in a fuse holder immediately below the main instrument panel and can be reached after releasing the cover, this being screwed in place. The fuses, rated at 10 amps, protect the various electrical circuits, these being marked adjacent to each fuse. A spare fuse is housed in the fuse holder.

2   The fuses are incorporated in the system to give protection from a sudden overload as could happen with a short circuit. If a fuse blows it should not be renewed until the cause of the short is found. This will involve checking the electrical circuit to correct the fault. If this rule is not observed, the fuse will almost certainly blow again.

3   When a fuse blows and no spare is available, a get-you-home remedy is to wrap the fuse in silver paper before replacing it in the fuse holder. The silver paper will restore electrical continuity by bridging the broken wire within the fuse. Replace the doctored fuse at the earliest opportunity to restore full circuit protection. Make sure any short circuit is eliminated first.

4   Always carry spare fuses of the correct rating.

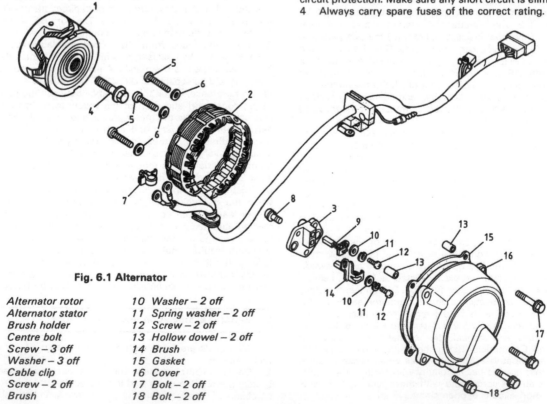

Fig. 6.1 Alternator

1 Alternator rotor	10 Washer – 2 off
2 Alternator stator	11 Spring washer – 2 off
3 Brush holder	12 Screw – 2 off
4 Centre bolt	13 Hollow dowel – 2 off
5 Screw – 3 off	14 Brush
6 Washer – 3 off	15 Gasket
7 Cable clip	16 Cover
8 Screw – 2 off	17 Bolt – 2 off
9 Brush	18 Bolt – 2 off

4.5 Regulator/rectifier unit is located here on CB650Z models

5.2a Battery is retained by this bar on CB650Z models

5.2b Maintain electrolyte level between upper and lower level lines

7.1a Main (30A) fuse is of the ribbon type and ...

7.1b ... fitted in this small box behind the right-hand side panel (650Z)

7.1c The remaining fuses are fitted as shown beneath their plastic cover

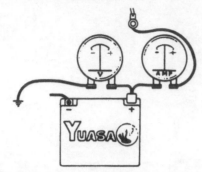

Fig. 6.2 Connecting ammeter and voltmeter for charging check

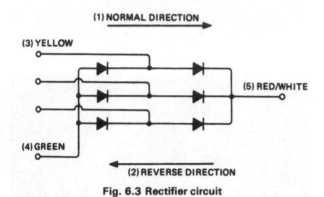

(1) NORMAL DIRECTION

(3) YELLOW

(5) RED/WHITE

(4) GREEN

(2) REVERSE DIRECTION

Fig. 6.3 Rectifier circuit

8 Starter motor: removal, examination and replacement

1   An electric starter motor, operated from a small pushbutton on the right-hand side of the handlebars, provides the necessary motive force for starting the engine. No kickstart lever is fitted. The starter motor is housed in a compartment to the rear of the cylinder block, beneath a chromed cover.

2   When the starter button is operated, a heavy duty solenoid relay is activated, switching the necessary heavy current directly from the battery to the motor. The motor turns the crankshaft, driving through an idler gear to the primary shaft mounted roller clutch. The clutch works on the ramp and roller principle; spring loaded rollers are forced into wedge-shaped slots to take up the drive. Thus, as soon as the engine starts, and the crankshaft starts to rotate more quickly than the driven clutch hub, the starter motor drive is automatically disconnected.

3   Before removing the motor, disconnect the positive (+) battery lead to isolate the electrical system. Release the starter motor cover by unscrewing the retaining bolts. Trace the heavy duty starter motor cable back to the solenoid and disconnect it after sliding back the protective boot. Release the starter motor mounting bolts, then remove the motor by drawing it to the left and then lifting it clear of its recess. The cable should be threaded clear of the frame.

4   The parts of the starter motor most likely to require attention are the brushes. The end cover is retained by the two long screws which pass through the lugs cast on both end pieces. If the screws are withdrawn, the end cover can be lifted away and the brush gear exposed.

5   Lift up the spring clips which bear on the end of each brush and remove the brushes from their holders. The standard length and wear limit of the brushes is given in the Specifications at the beginning of this Chapter. If either brush is worn to a length less than that of the wear limit, renew the brushes as a pair. When the brushes are badly worn, the starter motor will be unable to produce sufficient power to turn the engine over during starting.

6   Before the brushes are replaced, make sure the commutator on which they bear is clean. If necessary, the commutator segments may be burnished using Crokus paper. This is a fine abrasive paper produced for this purpose, and can be obtained from auto-electrical specialists. On no account use emery paper because fine particles from the paper may embed themselves in the soft commutator material, leading to rapid brush wear. After cleaning, wipe the commutator with a rag soaked in methylated spirits or de-natured alcohol to remove any dust or grease.

7   Honda do not supply undercut figures for the mica insulators which lie between each commutator segment, and thus imply that the armature assembly must be renewed when the commutator becomes worn. As this is a rather expensive solution, it may be worth seeking the advice of an auto-electrician who may be able to recondition the worn assembly at a favourable price.

8   Replace the brushes in their holders and check that they slide quite freely. Make sure the brushes are replaced in their original positions because they will have worn to the profile of the commutator. Replace and tighten the end cover, then replace the starter motor and cable in the housing, tighten down and re-make the electrical connection to the solenoid switch.

8.8 Refit the starter motor to its compartment

9 Starter solenoid switch: function and location

1   The starter motor switch is designed to work on the electro-magnetic principle. When the starter motor button is depressed, current from the battery passes through windings in the switch solenoid and generates an electro-magnetic force which causes a set of contact points to close. Immediately the points close, the starter motor is energised and a very heavy current is drawn from the battery.

2   This arrangement is used for two reasons. Firstly, the starter motor current is drawn only when the button is depressed and is cut off again when pressure on the button is released. This ensures minimum damage on the battery. Secondly, if the battery is in a low state of charge, there will not be sufficient current to cause the solenoid contacts to close. In consequence, it is not possible to place an excessive drain on the battery which, in some circumstances, can cause the plates to overheat and shed their coatings. If the starter will not operate, first suspect a discharged battery. This can be checked by trying the horn or switching on the lights. If this check shows the battery to be in good shape, suspect the starter switch which should come into action with a pronounced click. It is located behind

either the right-hand side panel (CB650 Z models) or the left-hand side panel (CB650 A and CA models), and can be identified by the heavy duty starter cable connected to it. It is not possible to effect a satisfactory repair if the switch malfunctions; it must be renewed.

3 Before condemning the starter solenoid, check that the malfunction is not caused by a defective clutch interlock switch or neutral interlock switch. Unless either or both of these are working, ie neutral has been selected and/or the clutch is disengaged, the starter will not operate. Check the operation of both switches as described later in this Chapter.

10 Headlamp: replacing bulbs

1 Depending on the model, the headlamp bulb is either held in a separate bulb holder, which fits into the rear of the reflector, or is an integral part of the reflector and therefore a sealed beam unit.

2 The UK market CB650 Z model has a separate bulb, of the quartz halogen type, with a 60/55W rating. The US market models are equipped with quartz halogen sealed beam units, with a rating of 65/50W.

3 In both cases initial access is gained by removing the headlamp rim, complete with glass and reflector, which is retained by screws through the headlamp shell. The socket at the rear of the reflector may be pulled off the terminal pins.

4 On models fitted with a separate headlamp bulb, prise off

the rubber boot which protects the bulb holder. The bulb is secured by two sprung arms which hinge from one side of the holder. Pinch the arms together to free them and then lift the bulb out. If the bulb has failed, it may now be discarded. If, however, the bulb is to be refitted, or a new replacement is to be installed, some caution must be exercised. As stated earlier, the bulb is of the halogen type with a quartz envelope. When removing or fitting the bulb on no account should the quartz envelope be handled. Use a piece of clean, dry cloth to insulate the bulb from the hand when removal or insertion is necessary. If the 'glass' is touched by the skin, any greasy or acidic deposits thereon will etch the quartz, leading to the formation of a 'hot spot'. Due to the high temperatures developed during the working of these bulbs, this will result in a weak point, and the bulb will fail prematurely. If the envelope is touched inadvertently, it should be cleaned with a solvent such as methylated spirits.

5 On UK models the parking lamp bulb holder is a press fit in the reflector, the bulb being of the bayonet fitting type. To remove the bulb, press it inwards and turn it anti-clockwise so that the bayonet pins disengage. This bulb has a rating of 4W.

6 If a sealed beam headlamp becomes defective it is necessary to renew the complete glass/reflector unit as the bulb element is not detachable. The unit is held by screws passing through tabs on the headlamp rim. It will also be necessary to disconnect the beam adjustment screw and tension spring which pass into a captive nut. Make an approximate note of the screw position to aid resetting the beam adjustment. No pilot bulb is fitted to the sealed beam unit.

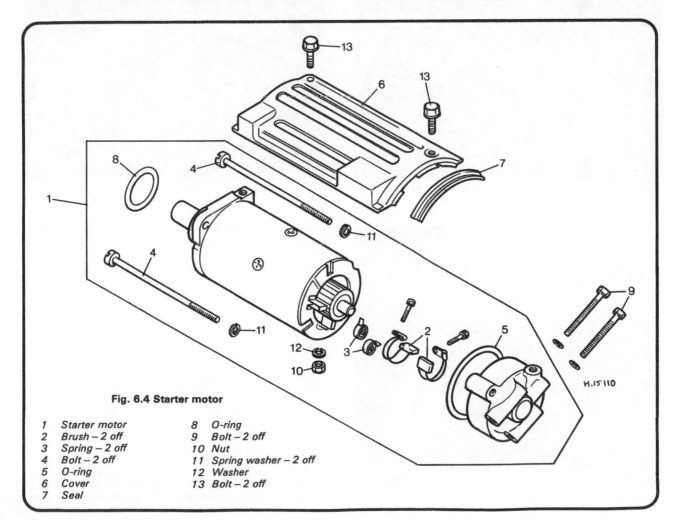

Fig. 6.4 Starter motor

1 Starter motor
2 Brush – 2 off
3 Spring – 2 off
4 Bolt – 2 off
5 O-ring
6 Cover
7 Seal
8 O-ring
9 Bolt – 2 off
10 Nut
11 Spring washer – 2 off
12 Washer
13 Bolt – 2 off

10.3a Remove screws to separate headlamp assembly from shell and ...

10.3b ... disconnect socket at rear of reflector

10.4a Bulb is retained by spring clip arrangement as shown

10.4b Pinch clip ends together and lift out bulb

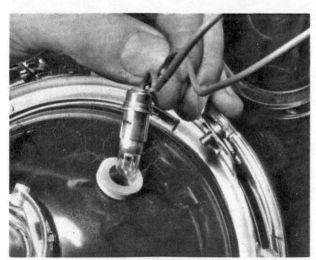

10.5 Pilot bulb holder is a press fit. Bulb is a bayonet fitting

11 Headlamp: adjusting beam height

1   Beam height is adjusted by fitting the headlamp about the two headlamp shell retaining bolts, after these have been slackened off. Horizontal adjustment is provided by a screw which passes into the headlamp rim.

2   To set up the headlamp the machine should be placed on level ground at least 25 feet away from a wall in its normal position, ie, off its stand and with the rider (and a pillion passenger if one is regularly carried) seated normally. On main beam the height of the centre of the headlamp beam from the ground, should be the same as the centre of the headlamp unit is from the ground. This will ensure that when the dip switch is operated the lamp will not give a dazzling effect.

3   Depending upon the country in question, the headlamp beam adjustment screw should be turned anti-clockwise (to redirect the beam to the right-hand side) or clockwise (to adjust the beam to the left-hand side).

4   UK lighting regulations stipulate that the lighting system must be arranged so that the light will not dazzle a person standing at a distance greater than 25 feet from the lamp, whose eye level is not less than 3ft 6 inches above that plane. It is easy to approximate this setting by placing the machine 25

feet away from a wall, on a level road and setting the beam height so that it is concentrated at the same height as the distance of the centre of the headlamp from the ground. The rider must be seated normally during this operation and also the pillion passenger, if one is carried regularly.

12 Stop and tail lamp: replacing bulbs – CB650 Z (UK) model

1   On this model, two twin filament stop/tail lamp bulbs are used. Each bulb is fitted to a separate detachable holder. To gain access to the holders, the dualseat must be raised and the tool compartment removed from the tail fairing. Both of the bulb holders pass through the rear wall of the tail fairing.

2   To remove a holder it should be twisted slightly, so that the locating lugs disengage from the aperture, and then withdrawn. Both bulbs are of the bayonet type, with offset pins to prevent incorrect positioning. The backs of the bulb holders are marked to show the correct way to twist them to lock them in position upon refitting.

11.1 This screw provides horizontal headlamp beam adjustment

12.1 Twin bulb rear light assembly on UK CB650Z model

12.2 Twist bulb holders to free them. Bulbs are bayonet fitting with offset pins

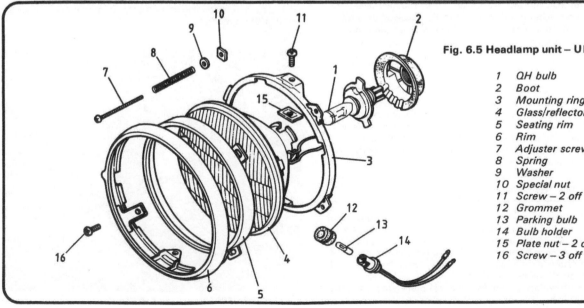

Fig. 6.5 Headlamp unit – UK model

1   QH bulb
2   Boot
3   Mounting ring
4   Glass/reflector unit
5   Seating rim
6   Rim
7   Adjuster screw
8   Spring
9   Washer
10 Special nut
11 Screw – 2 off
12 Grommet
13 Parking bulb
14 Bulb holder
15 Plate nut – 2 off
16 Screw – 3 off

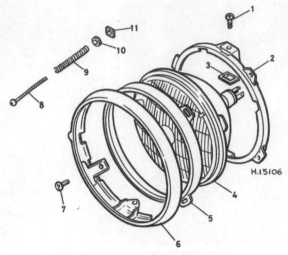

Fig. 6.6 Headlamp unit – US models

1	Screw – 2 off	7	Screw – 3 off
2	Mounting ring	8	Adjuster screw
3	Plate nut – 2 off	9	Spring
4	Sealed beam unit	10	Washer
5	Seating ring	11	Special nut
6	Rim		

13 Rear number plate lamp: replacing bulb – CB650 Z (UK) model

1   Rear number plate illumination is provided by a separate 10 watt bulb which fits into a holder directly between and below the two stop/tail lamp bulb holders. The holder is a push fit in the seat fairing wall and is secured by a strip of metal held by a single screw.

14 Rear lamp assembly: removal and examination – CB650 Z (UK) model

1   The rear lamp unit is unusually large and complicated on these machines, and forms part of the rear tail fairing. In the event of damage, a considerable amount of dismantling work is needed to remove the unit from the rear of the machine.
2   Start by releasing the rear grab handle, followed by the tail fairing assembly, the latter being retained by cross head screws. The rear lamp bracket is now exposed, as are the three bolts which pass through it to retain the rear lamp assembly. Release the bolts and lift the unit away, having removed the three bulbholders as described in Sections 12 and 13.
3   The large lens can now be removed by unscrewing the retaining screws which will have been exposed when the unit was removed from its mounting bracket. It will be noted that a double Fresnel lens diffuser is fitted between the red outer lens and the bulbs. This focuses the light from the twin bulbs, and must be in place if the lamp assembly is to work properly. The diffuser is retained by the three screws.

15 Rear lamp: replacing bulbs – US models

1   The rear lamp is a conventional unit mounted above the number (license) plate. The unit houses either a single twin-filament bulb (CB650 Z) or two twin-filament bulbs (CB650 A and CA). Bulb access is the same in both cases, by removing the lens. This is retained by four screws on the single bulb unit and two screws on the twin bulb lamp.
2   There is no separate number (license) plate lamp on these models.

16 Flashing indicator lamps: replacing bulbs

1   Flashing indicator lamps are fitted to the front and rear of the machine. They are mounted on short stalks through which the wires pass. Access to each bulb is gained by removing the plastic lens cover. The lens covers are retained by either two or three screws depending upon the model. Note that the three screws, which retain the lens covers to the angular units on the CB650 Z models, enter from the rear of the units.
2   All bulbs fitted to the rear units are of the single-filament type, as are the front bulbs of UK market models. Machines supplied to the US are fitted with double-filament bulbs at the front, with offset pins to prevent incorrect repositioning on replacement. The second filament is separately switched to allow the front indicators to serve as daytime running lamps.
3   To replace a bulb, remove the plastic lens cover by withdrawing the two or three retaining crosshead screws. Push the bulb in, turn it to the left and withdraw. Note it is important to use a correctly rated bulb otherwise the flashing rate will be altered. When refitting the lens cover, ensure the sealing ring is in good condition; a perished or split ring will lead to the ingress of moisture. Do not overtighten the two lens cover retaining screws on the CB650 A and CA models, or the cover will crack or fracture.

16.1 Lens cover retaining screws enter from rear of indicator unit (CB 650Z model)

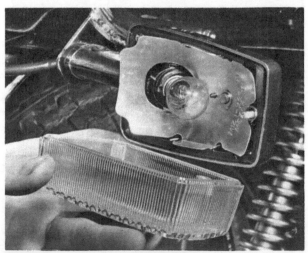

16.3 Remove lens cover to replace a bulb

17 Flasher unit: location and replacement

1   The flasher relay unit is located behind the left-hand side panel on the CB650 Z models and below the dualseat, to the rear of the battery, on the CB650 A and CA models. In each case, it is supported on an anti-vibration mounting made of rubber.

2   If the flasher unit is functioning correctly, a series of audible clicks will be heard when the indicator lamps are in action. If the unit malfunctions and all the bulbs are in working order, the usual symptom is one initial flash before the unit goes dead; it will be necessary to replace the unit complete if the fault cannot be attributed to any other cause.

3   Care should be taken when handling the unit to ensure that it is not dropped or otherwise subjected to shocks. The internal components may be irreparably damaged by such treatment.

18 Instrument illumination and warning panel lamps: replacing the bulbs

1   To gain access to the instrument and warning panel bulbs it will be necessary to dismantle the instrument panel as described in Chapter 4 or Chapter 7 (as applicable). This will allow access to the various instrument illumination and warning lamp bulbs. Check the recommended wattage with the Specification Section prior to renewal.

19 Horn: location and examination

1   The horn is suspended from a flexible steel strip bolted to the lower steering yoke behind the headlamp shroud.

2   A small screw and locknut arrangement provides adjustment. In the event that the horn's performance deteriorates significantly, experimentation with the screw setting will usually restore it to normal operation.

20 Ignition switch: removal and replacement

1   The combined ignition and lighting master switch is bolted to the upper yoke, and may be removed after the instrument panel has been detached. Disconnect the block connector plug from the base of the ignition switch. The switch is held in place by two screws, after the removal of which the switch can be displaced downwards.

2   Repair of a malfunctioning switch is not practicable as the component is a sealed unit, renewal is the only solution.

3   The switch may be refitted by reversing the dismantling sequence. Remember that a new switch will also require a new set of keys.

21 Stop lamp switch: adjustment

1   All models have a stop lamp switch fitted to operate in conjunction with the rear brake pedal. The switch is located immediately above the swinging arm on the right-hand side of the machine; it has a threaded body giving a range of adjustment.

2   If the stop lamp is late in operating, hold the body still and rotate the combined adjusting/mounting nut in an anti-clockwise direction so that the body moves away from the brake pedal shaft. If the switch operates too early or has a tendency to stick on, rotate the nut in a clockwise direction. As a guide, the light should operate after the brake pedal has been depressed by about 2 cm ($\frac{3}{4}$ inch). A stop lamp switch is also incorporated in the front brake system. The mechanical switch is a push fit in the handlebar lever stock. If the switch malfunctions, repair is impracticable. The switch should be renewed.

22 Handlebar switches: general

1   Generally speaking, the switches give little trouble but if necessary they can be dismantled by separating the halves which form a split clamp around the handlebars. Note that the machine cannot be started until the ignition cut-out on the right-hand end of the handlebars is turned to the centre 'ON' position.

2   Always disconnect the battery before removing any of the switches, to prevent the possibility of a short circuit. Most troubles are caused by dirty contacts, but in the event of the breakage of some internal part, it will be necessary to renew the complete switch.

3   Because the internal components of each switch are very small, and therefore difficult to dismantle and reassemble, it is suggested a special electrical contact cleaner be used to clean corroded contacts. This can be sprayed into each switch, without the need for dismantling.

23 Neutral switch: general

1   A small switch fitted to the left-hand side of the crankcase operates a warning lamp in the instrument panel to indicate the neutral has been selected. More importantly, it is inter-connected with the starter solenoid and will only allow the engine to be started if the gearbox is in neutral, unless the clutch is disengaged. It can be checked by setting a multimeter on the resistance scale and connecting one probe to the switch terminal and the other to earth. The meter should indicate continuity when neutral is selected and infinite resistance when in any gear.

24 Clutch interlock switch: general

1   A small plunger-type switch is incorporated in the clutch lever, serving to prevent operation of the starter circuit when any gear has been selected, unless the clutch lever is held in. It can be checked by the method described above for the neutral switch. If defective it must be renewed, as there is no satisfactory means of repair. The switch can be removed after releasing the clutch cable and lever blade.

25 Switch testing procedure

1   In the event of a suspected malfunction the various switch contacts can be checked for continuity in a similar manner to that described in the two preceding Sections. Details of the switch connections and the appropriate wiring will be found in the colour wiring diagrams that follow. Note that the electrical system should be isolated by disconnecting the battery leads to avoid short circuits.

17.2 Flasher unit will emit audible clicks if functioning correctly

18.1 With intruments separated from their bases, bulbs are accessible

21.2 Stop lamp switch is adjustable

22.1a Switch clusters clamp around handlebar ends and ...

22.1b ... may be separated for maintenance

26 Fault diagnosis: electrical system

Symptom	Cause	Remedy
Complete electrical failure	Blown fuse(s)	Check wiring and electrical components for short circuit before fitting new fuse
	Isolated battery	Check battery connections, also whether connections show signs of corrosion
Dim lights, horn and starter inoperatve	Discharged battery	Remove battery and charge with battery charger. Check generator output and voltage regulator/rectifier condition
Constantly blowing bulbs	Vibration or poor earth connection	Check security of bulb holders Check earth return connections
Parking lights dim rapidly	Battery will not hold charge	Renew battery at earliest opportunity
Tail lamp fails	Blown bulb or fuse	Renew
Headlamp fails	Blown bulb or fuse	Renew
Flashing indicators do not operate or flash fast or slow	Blown bulb Damaged flasher unit	Renew bulb Renew flasher unit
Horn inoperative or weak	Faulty switch Incorrect adjustment	Check switch Adjust
Incorrect charging	Faulty alternator Faulty rectifier Faulty regulator Wiring fault	Check Check Check Check
Over or under charging	As above, or faulty battery	Check
Starter motor sluggish	Worn brushes Dirty commutator	Remove starter motor and renew brushes Clean
Starter motor does not turn	Machine in gear Emergency switch in OFF position Faulty switches or wiring Battery float Loose battery terminal connections(s)	Disengage clutch Turn on Check continuity Recharge Check and tighten if necessary

Chapter 7 The 1981-on CB650, CB650 C and CB650 SC models

Contents

Specifications

Except where entered below, specifications for the models covered in this Chapter are the same as those given for the previous model at the beginning at each Chapter. Refer to Section 1 for details of model identification.

Specifications relating to Routine maintenance

Front fork oil capacity (per leg)
Dry:
CB650 B and C .... 210 cc (7.4 Imp fl oz, 7.1 US fl oz)
CB650 CB .... 245 cc (8.6 Imp fl oz, 8.3 US fl oz)
CB650 SC .... 320 cc (11.3 Imp fl oz, 10.8 US floz)
At oil change:
CB650 B and C .... 190 cc (6.7 Imp fl oz, 6.4 US fl oz)
CB650 CB .... 225 cc (7.9 Imp fl oz, 7.6 US fl oz)
CB650 SC .... 300 cc (10.6 Imp fl oz, 10.1 US fl oz)

Tyre pressures
Front .... 28 psi (2.0 kg/cm²)
Rear (solo):
CB650 B and CB .... 32 psi (2.25 kg/cm²)
CB650 C and SC .... 28 psi (2.0 kg/cm²)
Rear (with more than 90 kg/200 lb load):
CB650 B and C .... 36 psi (2.5 kg/cm²)
CB650 CB and SC .... 32 psi (2.25 kg/cm²)

Specifications relating to Chapter 1

Gearbox
Final drive ratio:
CB650 C .... 2.294 : 1 (17/39)
CB650 SC .... 2.235 : 1 (17/38)
Gear ratios – CB650 C and SC:
1st gear .... 2.737 : 1
2nd gear .... 1.778 : 1
3rd gear .... 1.333 : 1
4th gear .... 1.042 : 1
5th gear .... 0.885 : 1

Specifications relating to Chapter 2

Fuel tank – CB650 SC

Total capacity	13.5 litres (3.0 Imp gal, 3.5 US gal)
Reserve capacity	3.5 litres (0.8 Imp gal, 0.9 US gal)

Carburettors

	CB650 SC (UK)	CB650 SC (US)	All others (US)
Make	Keihin	Keihin	Keihin
Type	VB54 E	VB44 C	VB44 A
Main jet	95	118	120
Main air jet	150	80	100
Jet needle	12S	64A	64A
Pilot screw initial setting (turns out)	$2\frac{1}{8}$	$2\frac{1}{4}$	$2\frac{1}{4}$
Float height – all models	15.5 mm (0.61 in)		
Idle speed – all models	1050 ± 100 rpm		
Fast idle speed – all models	1000 – 2700 rpm		

Specifications relating to Chapter 3

Ignition timing

Retarded	10° BTDC @ 1050 rpm
Fully advanced	28° 30' BTDC @ 3600 rpm

Specifications relating to Chapter 4

Front forks

	Standard	Service limit
Fork spring free length:		
CB650 B and C	488.4 mm (19.23 in)	473.7 mm (18.65 in)
CB650 CB	559.6 mm (22.03 in)	542.7 mm (21.37 in)
CB650 SC (UK)	570.0 mm (22.44 in)	553.0 mm (21.77 in)
CB650 SC (US)	570.0 mm (22.44 in)	558.5 mm (21.99 in)
Stanchion OD:		
CB650 SC	36.950 – 36.975 mm (1.4547 – 1.4557 in)	
Service limit	36.90 mm (1.453 in)	
All other US models	34.975 – 34.950 mm (1.3770 – 1.3760 in)	
Service limit	34.90 mm (1.370 in)	
Lower leg ID:		
CB650 SC	38.040 – 38.080 mm (1.498 – 1.499 in)	
Service limit	38.20 mm (1.504 in)	
All other US models	Not available	
Bottom bush OD:		
CB650 SC	37.68 – 37.73 mm (1.483 – 1.485 in)	
Service limit	37.63 mm (1.482 in)	
All other US models	Not available	
Fork oil capacity	See Routine maintenance specifications in this Chapter	
Fork oil grade	Automatic transmission fluid (ATF)	

Rear suspension – CB650 SC

Travel	105 mm (4.1 in)
Spring free length	233.9 mm (9.21 in)
Service limit	229.0 mm (9.02 in)

Torque settings

	kgf m	lbf ft
Front fork air hose to top bolt	0.4 – 0.7	3.0 – 5.0
Front fork air hose adaptor to top bolt	0.4 – 0.7	3.0 – 5.0
Front fork air hose to adaptor	1.5 – 2.0	11.0 – 14.0
Front fork damper rod Allen bolt	1.5 – 2.5	11.0 – 18.0
Front fork top bolt	1.5 – 3.0	11.0 – 22.0

Specifications relating to Chapter 5

Wheels

Type:	
CB650 B and C	Wire-spoked
All other models	Comstar

Front brake

Type:	
CB650 B	Single hydraulic disc brake (single piston caliper)
CB650 C	Single hydraulic disc brake (twin piston caliper)
CB650 CB	Twin hydraulic disc brake (single piston calipers)
CB650 SC	Twin hydraulic disc brake (twin piston calipers)

Front brake (continued)
Caliper Bore ID:
CB650 C .... 30.230 – 30.280 mm (1.1902 – 1.1921 in)
Service limit.. 30.280 mm (1.193 in)
CB650 SC .... 27.000 – 27.076 mm (1.0630 – 1.0660 in)
Service limit.. 27.086 mm (1.066 in)
Caliper piston OD:
CB650 C .... 30.148 – 30.198 mm (1.1869 – 1.1889 in)
Service limit.. 30.140 mm (1.187 in)
CB650 SC .... 26.920 – 26.970 mm (1.0598 – 1.0618 in)
Service limit.. 26.912 mm (1.060 in)
Disc thickness:
CB650 C .... 6.8 – 7.2 mm (0.27 – 0.28 in)
Service limit.. 6.0 mm (0.24 in)
CB650 SC .... 4.8 – 5.2 mm (0.19 – 0.20 in)
Service limit.. 4.0 mm (0.16 in)

Tyres
Front:
CB650 B and C... 3.50H19 (4PR)
CB650 CB and SC.. 3.50S19 (4PR)
Rear:
CB650 B and C... 4.50H17 (4PR)
CB650 CB and SC.. 130/90S16

Tyre pressures .... See Routine maintenance specifications in this Chapter

Torque settings

	kgf m	lbf ft
Brake caliper body to mounting bracket bolts – CB650 C and SC	1.8 – 2.3	13.0 – 17.0
Front wheel spindle – CB650 C and SC	5.5 – 6.5	40.0 – 47.0
Front wheel spindle pinch bolt – CB650 C and SC	1.5 – 2.5	11.0 – 18.0
Brake hose union bolts – CB650 C and SC	2.5 – 3.5	18.0 – 25.0

1   Introduction

The first six Chapters of this manual relate to pre-1981 models. This Chapter describes the changes made to the later models, from 1981 onwards.

When working on 1981 or later model refer first to this Chapter. If the information required is not given in this Chapter it can be assumed that the task can be performed using the information given in the relevant Sections of Chapters 1 to 6.

To assist owners in identifying their machines correctly, the following text describes the modifications made from 1981 onwards.

1981 UK model
In 1981 the CB650 Z continued unchanged.

1982 UK models
In early 1982 the CB650 Z was discontinued and Honda introduced a new model the CB650 SC, also known as the Nighthawk. The CB650 SC was basically a custom version of the Z model with redesigned bodywork. Changes include the fitting of a 16 inch rear wheel, high handlebars, and a four-into-four exhaust system, with all components being similar to those already fitted on the US custom models. A black-painted engine unit was used to enhance the custom styling. The only major mechanical change related to the carburettors, the old slide type carburettors having been replaced with a set of constant depression units. Chassis changes included the fitting of twin piston calipers to the front brake and modified, air-assisted, front forks. The CB650 SC continued unchanged until it was discontinued in early 1984.

1981 US models
Along with the usual colour and graphics changes the 1981 CB650 B and CB models were also equipped with a modified 'air-assisted' front fork, and new carburettors, the old slide type having replaced by a set of constant depression units.

On the CB650 CB the only other major change was the fitting of a twin disc front brake to replace the old single disc set-up.

1982 US models
The only major difference between the 1982 CB650 C and the 1981 B model was that the single piston front brake caliper was replaced with a twin piston caliper.

In early 1982 the CB650 CB model was superseded by the CB650 SC, also known as the Nighthawk. The CB650 SC was fitted with redesigned bodywork and a new colour scheme to update the styling. The engine was also black-painted. The only technical change was that the front brake was uprated with the fitting of twin piston calipers.

Both the CB650 C and SC models were discontinued at the end of 1982.

2   Routine maintenance: service schedules

Note that revised service schedules have been introduced for all later models. Routine maintenance should now be carried out at the intervals given under the relevant sub-heading. Unless otherwise stated, the task can be performed using the information in the main Routine Maintenance section at the start of this manual, referring also to the revised Specifications at the start of this Chapter.

All models
Daily (pre-ride) check
Check the engine oil level
Check the fuel level. Ensure you have enough fuel to complete your journey or at least to get you to the nearest filling station
Check the wheels and tyres, especially tyre pressures and tread wear
Check that all controls are correctly adjusted and working smoothly
Check that the speedometer, horn and lights are working correctly

UK models
Every 600 miles (1000 km)
Repeat all the tasks listed under the daily checks and:
   Clean, lubricate and adjust the final drive chain

Monthly, or every 3600 miles (6000 km)
Repeat all previous maintenance tasks and check the following:
   Check the battery electrolyte level
   Check the front brake fluid level

Every 3600 miles (6000 km)
Carry out all tasks listed under the previous headings then carry out the following:
   Change the engine oil and filter
   Check and adjust the valve clearances
   Check and adjust the cam chain tension
   Check the spark plug condition
   Clean the air filter
   Synchronise and adjust the carburettors – refer to Section 3
   Inspect the fuel pipe for signs of damage – refer to Chapter 2
   Check the brake pad/shoe wear – refer to Section 3
   Check and adjust the clutch
   Check and lubricate the side stand pivot
   Clean the fuel tap bowl filter – CB650 SC only – refer to Section 3
   Drain the crankcase breather tube – CB650 SC only
   Check the wheels – refer to Chapter 5
   Check all fasteners for tightness

Every 7200 miles (12 000 km)
Repeat all the tasks given under the previous headings, then carry out the following:
   Renew the spark plugs
   Renew the air filter
   Check the operation of the steering and suspension and lubricate the swinging arm bushes – refer to Chapter 4, Routine maintenance and Section 3 of this Chapter for further information

Two yearly or every 10 800 miles (18 000 km)
   Renew the front brake fluid

US models
Monthly or every 300 miles (500 km)
Repeat all the tasks listed under the daily heading and:
   Clean, lubricate and adjust the final drive chain

Monthly or every 4000 miles (6400 km)
Repeat all the tasks listed under the previous headings and carry out the following:
   Check the battery electrolyte level
   Check the front brake fluid level

Every 4000 miles (6400 km)
Carry out all the tasks listed under previous checks, then carry out the following:
   Change the engine oil and filter
   Check and adjust the valve clearances
   Check and adjust the cam chain tension
   Renew the spark plugs
   Clean the air filter
   Synchronise and adjust the carburettors – refer to Section 3
   Inspect the fuel pipe for signs of damage – refer to Chapter 2
   Check the brake pad/shoe wear – refer to Section 3
   Check and adjust the clutch
   Clean the fuel tap bowl filter (1982 models only) – refer to Section 3
   Check and lubricate the side stand pivot
   Drain the crankcase breather tube
   Check the wheels – refer to Chapter 5
   Check all fasteners for tightness

Every 8000 miles (12 800 km)
Repeat all the tasks given under the previous headings and carry out the following:
   Renew the air filter
   Check the operation of the steering and suspension and

lubricate the swinging arm bushes – refer to Chapter 4, Routine maintenance and Section 3 of this Chapter for further information

Two yearly or every 12 000 miles (19 200 km)
   Renew the front brake fluid

---

3   Routine maintenance: new and revised procedures

Carburettors: synchronisation – constant depression carburettors
1   Constant depression carburettors can be synchronised using the information given in Routine maintenance noting that the adjusting screws and locknuts are situated in the throttle linkage and that the carburettor tops must be left in place. Use of the service tool, Part Number 07908-4600200, will make adjustment easier although this is not essential.

Fuel tap bowl filter: cleaning – CB650 C and SC models
Note: *Petrol (gasoline) is extremely flammable, especially when in the form of vapour. Take all precautions to prevent the risk of fire and read the Safety first! Section of this manual before starting work.*
2   Ensure the fuel tap is set to the OFF position and unscrew the filter bowl from the bottom of the tap. Be prepared to catch the small amount of fuel which will be released and mop up any spilt fuel immediately. Carefully prise out the O-ring from the base of the tap and remove the filter.
3   Wash both the filter and bowl in clean high flash-point solvent and examine the filter for cracks and splits or signs of clogging, renewing it if necessary. Renew the O-ring regardless of its apparent condition.
4   On refitting, ensure the index mark on the edge of the filter aligns with the index mark on the inside of the fuel tap and fit the new O-ring. Refit the bowl to the tap and use a close-fitting ring spanner on its cast hexagon to tighten it lightly; if possible tighten it to the specified torque setting of 0.3 – 0.5 kgf m (26 – 38 lbf in). Turn the tap to the ON position and thoroughly check for fuel leaks before taking the machine on the road.

Brake pads: checking for wear
5   On all 1981 models (single piston caliper) the brake pads can be checked and renewed as described in Routine maintenance.
6   On 1982-on (twin piston caliper) models the brake pad wear is checked in a similar manner using the wear line on the side of each pad's friction material. This line is visible on the upper edge of the inner brake pad. If the wear line is in contact with, or very close to, the disc the pads should be renewed as a set as follows.

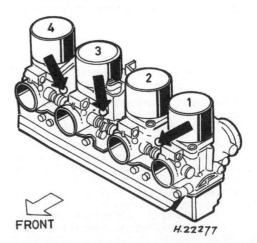

Fig. 7.1 Carburettor synchronising screw positions (arrowed) – CB650 SC (UK) and all US models (Sec 3)

7   To remove the pads first remove the bolt which retains the pad pin retaining plate and remove the plate from the back of the caliper. Using hand pressure, push the caliper towards the disc so that the pistons are pushed back into their bores. This is necessary to gain the clearance needed for the extra thickness of the new pads. While pushing the caliper in, maintain a careful watch on the fluid in the handlebar reservoir. If the reservoir has been overfilled, the surplus fluid will prevent the pistons returning fully and must be removed by soaking it up with a clean rag, taking great care to prevent fluid spillage. Remove the mounting bolt which secures the caliper body to its mounting bracket and swing the caliper away from the disc by pivoting it upwards about its top mounting. Withdraw both pad retaining pins and remove the pads from the caliper.

8   Brake pads are installed by a reverse of the removal procedure. Position the pads in the caliper and insert the pad retaining pins. Swing the caliper back down into position and refit its mounting bolt, tightening it to the specified torque setting. Refit the pad pin retaining plate, ensuring it locates correctly with the pins, and secure it in position with its retaining bolt.

9   Apply the brake lever gently and repeatedly to bring the pads firmly into contact with the disc and until full brake pressure is restored. Check the fluid level in the reservoir as described in Routine maintenance and top up if necessary. Before taking the machine out on the road, carefully check for fluid leaks from the system, and check that the front brake is working correctly. Remember also that new pads, and to a lesser extent cleaned pads, will require a bedding in period before they will function at peak efficiency. Where new pads are fitted use the brake firmly but gently for the first 50 – 100 miles to enable the pads to bed in fully.

Suspension: checking the front fork air pressure – CB650 SC (UK) and all US models

Note: *Do not use a tyre pressure gauge to check the air pressure; they are not accurate enough and lose too much air when disconnected. Never use an air line to add air to the forks, because they operate at far too high a pressure and could damage the seals. It is recommended that only a bicycle pump or one of the specialist aftermarket kits is used to add air; the latter usually comes equipped with its own built in gauge and is extremely accurate in use.*

10   The fork legs are linked by a balance pipe and are adjusted via the valve in the right-hand fork top bolt. The fork air pressure must be checked with the machine on its centre stand and with the front wheel clear of the ground to prevent the pressure being artificially increased due to the weight of the machine. This can be achieved by placing a suitable block underneath the engine unit. The specified pressure range is 0.7 – 1.1 kg/cm^2 (10 – 16 psi). On no account should the pressure in the forks ever exceed 3 kg/cm^2 (43 psi) as this will almost certainly damage the fork seals.

11   Always refit the valve's dust cap after checking/adjusting the air pressure.

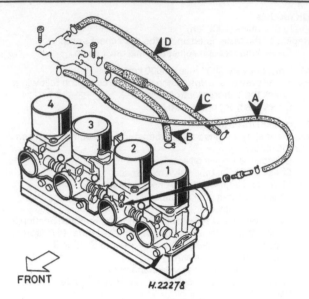

FRONT   H.22278

Fig. 7.2 Carburettor vacuum-operated valve – 1981-on US models (Sec 4)

A   Vacuum pipe to No 2 carburettor
B   Fuel feed pipe to carburettors
C   Fuel pipe from tank
D   Air vent pipe

diaphragm. If a leak develops in either of these the valve will not open and no fuel will flow to the carburettors. Inspect the valve as follows.

3   Turn the fuel tap on the tank to the OFF position and disconnect the fuel pipe from the tap. Remove the fuel tank as described in Chapter 2. Check the valve's vacuum pipe for signs of cracking or splitting, renewing it if necessary. If the pipe is in good condition, disconnect it from the number 2 carburettor intake tract and also disconnect the fuel feed pipe to the carburettors. Slacken the mounting screws and remove the valve from the carburettors. Reconnect the fuel pipe at the tap, turn the tap to the ON position, and suck gently on the end of the vacuum pipe. As this is done, fuel should flow out of the carburettor feed pipe. If not, the fuel valve is proven faulty and must be renewed; repairs are not possible.

4   On refitting, ensure the fuel valve pipes are securely fitted in their original positions and secured by their wire clips. Tighten the valve's mounting screws securely. Refit the fuel tank as described in Chapter 2 and check for leaks before taking the machine on the road.

4   Fuel system: modification – 1981 on US models

Note: *Gasoline (petrol) is extremely flammable, especially in the form of vapour. Take all precautions to prevent the risk of fire and read the Safety first! section of this manual before starting work.*

1   On 1981-on US models a vacuum-operated valve is fitted to the feed pipe between the tank and carburettors to control the flow of fuel. This valve is mounted on the top of number 3 and 4 carburettors, to the rear of the carburettor tops. When the engine is stopped the spring-loaded diaphragm in the valve is held in the closed position to prevent the flow of fuel. When the engine is started, the low pressure in the intake tract (of number 2 carburettor) opens up the diaphragm allowing fuel to flow to the carburettors.

2   In the event of failure, the most likely culprits are the vacuum pipe connecting the valve to number 2 carburettor, or the

5   Carburettors: separation and rejoining – CB650 SC (UK) and all US models

Note: *Petrol (gasoline) is extremely flammable, especially in the form of vapour. Take all precautions to prevent the risk of fire and read the Safety first! section of this manual before starting work.*

1   Never remove the carburettors from their mounting brackets unless absolutely necessary; each carburettor can be dismantled sufficiently for all normal cleaning and adjustment while in place on the mounting brackets. If separation is necessary it should be noted that Honda recommend that the choke valves, shafts and screws be renewed if disturbed.

2   Remove the carburettors as described in Chapter 2 and mark each carburettor body with its respective cylinder number. Unhook the choke relief spring from its groove in the choke shaft, situated between number 2 and 3 carburettors. Slacken the three

throttle linkage synchronising screw locknuts and screw each screw fully in until it seats, counting and recording the number of turns necessary to do so. This figure should be recorded, to be used later when rejoining and synchronising the carburettors.

3   Remove the screw which secures the fuel joint mounting plate to the rear mounting bracket and remove the plate. Slacken all the front and rear mounting bracket screws and remove both brackets from the carburettors. Once both brackets have been removed disengage the throttle linkage and carefully separate number 2 and 3 carburettors. Remove the air, fuel and accelerator pump joints noting the O-rings fitted on the fuel and accelerator pump joints.

4   If it is necessary to separate numbers 1 and 2 and/or 3 and 4 carburettors, it is first necessary to remove the choke valves and shaft as follows. Carefully file the staked ends off of the choke valve screws whilst taking great care not to allow any of the filings to enter the carburettor bore. Slacken all the choke screws and remove the valves from the carburettors. Note, before removing the choke valves from number 1 and 2 carburettors note the correct position of the choke relief and disengage it from the choke linkage. Withdraw the choke shaft from the carburettor body and discard all the choke components; new items must be fitted on reassembly. Once the choke shafts have been removed disengage the throttle linkage and carefully separate the individual carburettors. Note the spring fitted to the throttle linkage and the rubber boot fitted to the choke shaft between each set of carburettors. Remove the air, fuel and accelerator pump joints.

5   The carburettors are rejoined by a direct reversal of the separation sequence. Renew all fuel and accelerator pump joint O-rings regardless of their apparent condition and lubricate them with a smear of engine oil to aid reassembly.

6   First reassemble the two pairs of carburettors using their previously marked cylinder numbers to ensure they are correctly positioned. Fit the air, fuel and accelerator pump joints to number 1 carburettor and carefully rejoin it with the number 2 carburettor, ensuring that all joints are correctly aligned. Engage the throttle linkage arms ensuring that the forked arm is located between the two plain washers and not directly against the spring or screw. Fit the relief spring to the new choke shaft and insert the shaft into the carburettors, not omitting the rubber boot which is fitted between the carburettors. Ensure the relief spring is correctly located with the choke linkage and install the choke valves, tab washers and retaining bolts. Tighten the choke valve retaining bolts finger-tight only. Repeat the above sequence for numbers 3 and 4 carburettors.

7   Fit the air, fuel and accelerator pump joints to number 2 carburettor and carefully rejoin the two pairs of carburettors. Engage the throttle linkage as described above and refit the front mounting bracket, tightening its retaining screws loosely. Place the carburettors on a flat surface to ensure that they are correctly aligned, then tighten the mounting screws evenly and progressively until all are securely tightened. Refit the rear mounting bracket in a similar manner.

8   Refit the throttle linkage springs between number 1 and 2, and 3 and 4 carburettors. Reset the throttle linkage synchronising screws by screwing them fully in and then backing them out to their previously recorded positions. Check that the distance between each of the throttle valves and the pilot by-pass hole is equal and adjust as necessary. Tighten the synchronising screw locknuts securely and check that the throttle linkage operates smoothly and returns quickly. If not, slacken the carburettor mounting bracket screws and realign the carburettors as described in the preceding paragraph.

9   Engage the choke relief spring with the choke shaft of number 3 carburettor and close the choke valves by turning the choke linkage. With the choke valves closed tighten all their retaining bolts securely and then release the choke linkage making sure it returns smoothly. If all is well, close up the choke valves again and secure the retaining bolts in position by bending up the tab washers against the flats of each bolt.

10   Refit the carburettor assembly to the machine as described in Chapter 2. Although the synchronisation has been roughly set up, it will be necessary to carry out carburettor synchronisation

as described in Section 3 of this Chapter and Routine Maintenance.

6   Carburettors: dismantling, examination and reassembly – CB650 SC (UK) and all US models

Note: *Petrol (gasoline) is extremely flammable, especially in the form of vapour. Take all precautions to prevent the risk of fire and read the Safety first! section of this manual before starting work.*

1   Remove the carburettors from the machine as described in Chapter 2. Never remove the carburettors from their mounting brackets unless absolutely necessary; each carburettor can be dismantled sufficiently for all normal cleaning or adjustment procedures whilst in position on the brackets.

2   Working on one carburettor at a time to avoid the accidental interchange of components, slacken and remove the two screws which retain the carburettor top. Lift off the top, remove the spring from inside the piston and withdraw the piston from the carburettor. Remove the blanking plug from the top of the piston and unscrew the jet needle retainer. The jet needle can then be tipped out of the piston. Carefully prise out the rubber sealing ring from the top of the carburettor and remove the air jet cover which is retained by a single screw.

3   Remove the three screws which retain the float chamber to the bottom of the carburettor and lift off the chamber to gain access to the various jets. Press out the float pivot pin and remove the float and needle valve assembly.

4   The main jet is screwed into the bottom of the needle jet holder and can be removed with a flat-bladed screwdriver. The needle jet holder is screwed into the carburettor body and can be removed with a suitable spanner. The needle jet can then be pushed out of position from the top of the carburettor with a wooden dowel. The pilot jet is situated next to the needle jet holder. On CB650 C and SC models the jet can be unscrewed and removed from the carburettor, but on all other models it is an integral part of the carburettor body and cannot be removed.

5   The pilot air (mixture) screw is situated in the bottom of the carburettor body, in front of the float chamber. US models have a limiter cap fitted which prevents unwarranted adjustment of the mixture by providing only very limited adjustment of the screw with the float chamber in situ. On all models screw the pilot screw in until it seats lightly, counting the number of turns necessary to do so, then remove the screw along with its spring, flat washer and O-ring. The screw must be renewed if bent or damaged in any way.

6   The choke valves can be removed and renewed, if necessary, as described in the preceding Section of this Chapter.

7   Carefully check the carburettor body, float chamber and carburettor top for such damage as cracks, splits or distorted sealing faces. Whilst it may be necessary to repair such defects it will usually be necessary to renew the affected component.

8   Check that the piston and carburettor bore are not scored or worn. Damage of any sort will require the renewal of both components. Check that the needle is straight by rolling it on a flat surface such as a sheet of plate glass. If it is bent it must be renewed as a set along with the needle jet and holder.

9   Check that the floats are in good order and are not punctured. If either float is punctured it will produce the wrong fuel level in the float chamber, leading to an over-rich mixture and flooding. If the floats are damaged they must be renewed as satisfactory repair is not possible.

10   The float needle valve and seat will wear after lengthy service and should be closely examined, with a magnifying glass if necessary. Wear will usually take place in the form of a ridge or groove, which will cause the needle to seat imperfectly. Test the spring-loaded tip on the bottom of the needle valve by pushing it into the body of the needle. The tip should return quickly and easily under spring pressure. If the needle valve or seat are damaged both should be renewed as a set. However, the needle

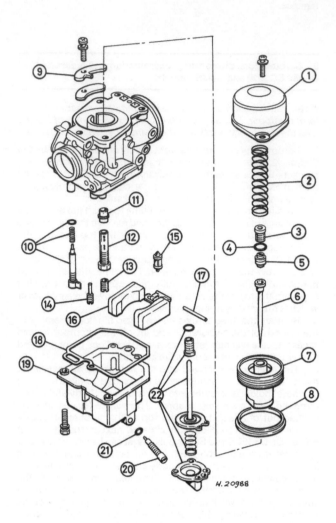

strong light; renew it if any are found. On refitting ensure the diaphragm is correctly seated and refit the spring and cover, tightening its retaining screws securely.

13   Before the carburettor is reassembled, using a reverse of the dismantling procedure, it should be cleaned out thoroughly, preferably by the use of compressed air. Avoid using a rag because there is always risk of fine particles of lint obstructing the internal air passages or the jet orifices. Check carefully the condition of all O-rings and gaskets and renew any that are worn or distorted. Never use a piece of wire or sharp metal object to clear a blocked jet. It is all too easy to enlarge the jet under these circumstances and increase the rate of fuel consumption. Always use compressed air to clear a blockage; a tyre pump makes an admirable substitute when a compressed air line is not available. Do not use excessive force when reassembling the carburettor because it is quite easy to shear the small jets or some of the smaller screws.

14   If the pilot air (mixture) screws are being re-used, reset them to their original positions as noted during dismantling. If new pilot screws are being fitted, set the screws to their specified initial setting, given in the Specifications at the start of this Chapter, noting that it will be necessary to adjust them as described in the following Section. Note, on US models do not install the limiter caps until the pilot screws have been adjusted. When refitting the pilot jets on CB650 C and SC models, gently screw the jet in until it seats lightly, then tighten it a further ¾ of a turn more.

15   Before refitting the carburettors to the machine check the float height and fast idle and accelerator pump adjustment as described in the following Section.

7   Carburettors: adjustments – CB650 SC (UK) and all US models

Note: *Petrol (gasoline) is extremely flammable, especially in the form of vapour. Take all precautions to prevent the risk of fire and read the Safety first! section of this manual before starting work.*

1   The first step in carburettor adjustment is to ensure that the jet sizes and float height are correct, which will require the removal and disassembly of the carburettors as described in Section 6.

2   Before any dismantling is undertaken eliminate all other possible causes of running problems, checking in particular the spark plugs, air filter element and the operation of the choke mechanism. Checking and cleaning these items will often resolve a mysterious flat spot or misfire.

Float height check

3   If the carburettors have been removed for the purpose of checking jet sizes, the float height should be measured at the same time. It is unlikely that once set up correctly, there will be any significant variation, unless the float needle or seat have worn. These should be checked and renewed as required. With the float chamber removed, slowly rotate the carburettors until gravity acting on the floats moves the float until the valve is just closed, but not so far that the needle's spring-loaded tip is compressed. Measure the distance between the gasket face and the bottom of the float with an accurate ruler. The correct setting should be as given in the Specifications Section. If the float height is not correct, the float must be renewed as adjustments are not possible. Repeat the procedure on the other carburettors.

Accelerator pump and fast idle adjustment

4   Once the float heights are known to be correct refit the float chambers and check the accelerator pump and fast idle adjustment as follows. (Refer to Chapter 2 for a general description of accelerator pump and fast idle mechanism operation.)

5   To check that the accelerator pump (fitted to number 2 carburettor) is correctly adjusted, ensure the throttle valves are completely closed and measure the clearance between the tip of

Fig. 7.3 Carburettor component parts – CB650 SC (UK) and all US models (Sec 6)

1 Carburettor top	12 Needle jet holder
2 Spring	13 Main jet
3 Blanking plug	14 Pilot jet
4 O-ring	15 Needle valve
5 Jet needle retainer	16 Float
6 Jet needle	17 Pivot pin
7 Piston	18 O-ring
8 Sealing ring	19 Float chamber
9 Air jet cover	20 Drain screw
10 Pilot air (mixture)	21 O-ring
screw	22 Accelerator pump (No 2
11 Needle jet	carburettor only)

valve seat is an integral part of the carburettor body and is not available separately.

11   The slow and main air jets, and on some models the pilot jet, are fixed into the carburettor body. Neither jet can be removed but all should be examined to make sure that they are clean and free from obstructions.

12   An accelerator pump is fitted to the float chamber of number 2 carburettor. Remove the pump cover retaining screws, lift off the cover and spring, and remove the diaphragm assembly. Inspect the diaphragm for perforations by holding it up to a

the accelerator pump rod and its operating tang. This distance should be 0 – 0.04 mm (0 – 0.0016 in). If this is not the case, adjust it by carefully bending the operating tang until the correct clearance is obtained.

6   The fast idle mechanism adjustment is checked with the throttle valves fully closed and the choke valves fully open. In this position there should be 0.7 – 1.0 mm (0.03 – 0.04 in) clearance between the throttle linkage and the fast idle arm adjusting pin. If not, this can be adjusted by carefully opening and closing the forked end of the fast idle adjusting arm.

Pilot air (mixture) screw adjustment

7   If the pilot screws have been removed or disturbed they must be adjusted as follows. *On US models do not adjust the pilot screws unnecessarily; they are preset at the factory and fitted with a limiter cap to prevent unnecessary adjustment.* Turn each pilot screw in until it seats lightly, then unscrew it by the number of turns shown in the Specifications at the start of this Chapter. This is only an initial setting and the pilot screws must then be adjusted as described below.

8   Start the engine, warming it up to normal operating temperature, then set the machine to its specified idle speed by rotating the throttle stop screw situated on the underside of the carburettors. With the engine running at its specified idle speed turn all the pilot screws out a further half a turn whilst noting the effect on the engine speed. If the engine speed increases by 50 rpm or more turn all the screws out a further half a turn. Repeat this procedure until the engine speed fails to rise, then reset the engine to the specified idle speed using the throttle stop screw. Once the idle speed is correct, slowly turn the pilot screw of number 1 in whilst noting the effect this has on idle speed. Once the idle speed has dropped 50 rpm, stop and back the screw off by one complete turn, then reset the engine to the specified idle speed. Repeat this procedure on the remaining three pilot screws.

9   On US models once the pilot screws are correctly set it is necessary to fit a new limiter cap to each screw. The cap must be glued over the pilot screw using Loctite 601 or a suitable equivalent whilst taking care not to move the screw. The cap must be positioned so that its lug is tight against the stop on the float chamber, so preventing the screw being turned anti-clockwise and richening the mixture, yet permitting a small amount of movement clockwise.

10   On all models, ensure that the engine idles smoothly and does not falter and stop after the throttle twistgrip has been opened and closed a few times. Finally adjust the throttle cables and check that the throttle valve operates smoothly and returns quickly before taking the machine on the road.

---

8   Front forks: removal and refitting – CB650 SC (UK) and all US models

---

1   Place the machine on its centre stand and raise the front wheel clear of the ground by placing a suitable block beneath the engine unit.

2   Remove the front wheel as described in either Chapter 5 (CB650 B and C models) or in Section 12 of this Chapter (CB650 CB and SC models). On CB650 CB and SC models remove the remaining brake caliper from the fork leg and tie it to the frame. On all models, slacken the four bolts which secure the mudguard to the forks and remove it from the machine.

3   Unscrew the air valve dust cap from the top of the right-hand fork leg and depress the valve to release any air pressure from the fork legs. Slacken the two screws which retain the fusebox cover to the top yoke and lift off the cover. Unscrew the air pressure connecting hose from both the left and right-hand fork top bolts and remove it from the machine. Once the hose has been removed unscrew the air hose adaptor from the right-hand top bolt.

4   Slacken the top and bottom yoke pinch bolts and remove the

forks by twisting them and pulling downwards. If the fork legs are seized in the yokes, spray the area with penetrating oil and allow time for it to work. In severe cases the split clamps can be sprung apart with a large flat-bladed screwdriver, although extreme care must be taken not to distort or fracture the casting. Note that if the fork legs are to be dismantled it is preferable to slacken their top bolts whilst they are still held in the yokes.

5   The forks are refitted in a reverse of the removal procedure noting the following points. Slide the forks back into position in the yokes so that the top surface of each stanchion is flush with the yoke and the top bolt air hose unions are correctly positioned. Tighten both the top and bottom yoke pinch bolts securely. If the fork legs are being refitted after an overhaul, tighten the top bolts to their specified torque settings now that they are securely held in the yokes.

6   Fit new O-rings to the air hose adaptor and the left-hand end of the air hose and smear both the O-rings with a small amount of grease. Refit the adaptor to the right-hand top bolt and install the air hose, tightening all connections to their specified torque settings.

7   Refit the mudguard, tightening its retaining bolts securely, and refit the front wheel as described in either Chapter 5 or Section 12 of this Chapter (as applicable). Charge the forks with the specified amount of air, as described in Section 3, and refit the dust cap on the valve. Thoroughly check the operation of the front forks before taking the machine on the road.

---

9   Front forks: dismantling and reassembly – CB650 SC (UK) and all US models

---

1   Remove the fork legs from the machine as described in the previous Section. Dismantle each leg separately to avoid interchanging parts.

2   Clamp the stanchion in a vice equipped with soft jaws, being careful not to overtighten or score the surface of the stanchion. Extend the fork leg fully and unscrew the top bolt from the stanchion, whilst taking care not to allow it to be forcibly expelled by spring pressure as the last threads of the bolt are unscrewed. Withdraw the spring from the stanchion.

3   Invert the fork leg over a suitable container and vigorously pump the leg to remove as much oil as possible.

4   Once the oil has drained, slacken the Allen-headed bolt which passes up through the bottom of the lower leg and into the damper rod. It is quite likely that the damper rod will rotate in the lower leg instead of unscrewing and thus impede the removal of the bolt. If this problem arises, temporarily refit the fork spring and top bolt to place pressure on damper rod's head. If the damper rod still rotates, remove the top bolt and spring, and obtain a length of wooden dowel of sufficient dimensions to pass down through the stanchion to engage in the damper rod's head, and to extend at least 6 inches from the top of the stanchion. Chamfer the lower end of the dowel to fit into the damper rod. Have an assistant hold the dowel firmly in place whilst the Allen bolt is slackened.

5   Once the damper rod bolt has been removed carefully lever the dust seal out of the top of the lower leg and remove the circlip and washer. The stanchion and lower leg can then be separated by pulling the two components apart.

6   Push the stanchion fully into the lower leg, taking care not to damage the damper rod seat, and pull it out as sharply as possible. Repeat the operation several times, using the slide-hammer action of the bottom bush against the top bush to dislodge the oil seal, until the stanchion is freed from the lower leg. Invert the stanchion and tip out the damper rod and rebound spring. The oil seal, special washer and top bush can then be slid off the stanchion. Removal of the lower bush should only be attempted if renewal is required; it can be removed by inserting a screwdriver into the vertical split and levering the two ends apart by just enough to allow the bush to be slid off the bottom of the stanchion. Invert the lower leg and tip out the damper rod seat.

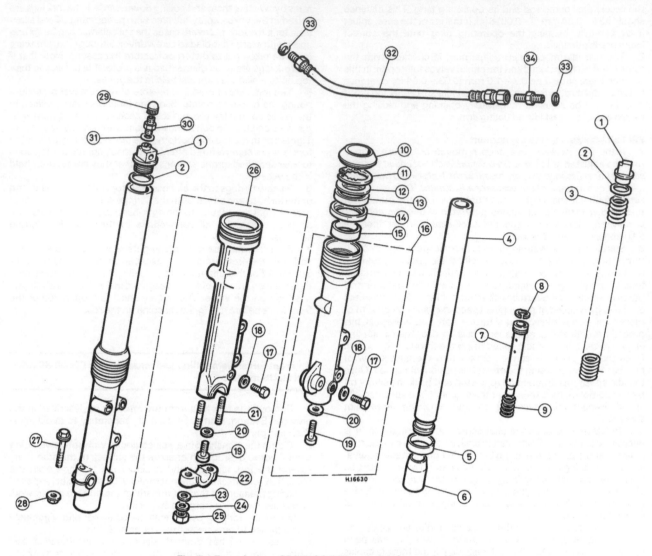

Fig. 7.4 Front forks – CB650 SC (UK) and all US models (Sec 9)

1 Top bolt	10 Dust seal	19 Allen bolt	28 Nut
2 O-ring	11 Circlip	20 Sealing washer	29 Dust cap
3 Spring	12 Washer	21 Stud – 2 off	30 Air valve
4 Stanchion	13 Oil seal	22 Spindle clamp	31 O-ring
5 Bottom bush	14 Special washer	23 Washer – 2 off	32 Air hose
6 Damper rod seat	15 Top bush	24 Spring washer – 2 off	33 O-ring – 2 off
7 Damper rod	16 Lower leg – CB650 B and C models	25 Nut – 2 off	34 Union
8 Piston ring	17 Oil drain bolt	26 Lower leg – CB650 CB and SC models	
9 Rebound spring	18 Sealing washer	27 Pinch bolt	

7   When all components have been cleaned and examined as described in the following Section, they should be reassembled in the reverse of the removal sequence.

8   Refit the rebound spring to the damper rod and insert it into the upper end off the stanchion. Use the fork spring to push the damper rod down until it projects fully from the lower end of the stanchion and refit the damper rod seat to the end of the rod. Refit the bottom bush (if removed) to the lower end of the stanchion, ensuring that it is correctly positioned in its locating groove.

9   Smear the bottom bush and stanchion with oil (ATF) and insert the assembly into the lower leg. Press the stanchion fully into the lower leg to centralise the damper rod seat. Check the condition of the sealing washer on the Allen bolt and renew it if necessary. Apply a few drops of thread-locking compound to the threads of the Allen bolt and install it in the lower leg. Tighten the bolt to the specified torque setting whilst, if necessary, holding the damper rod head using the method employed on dismantling.

10   Lubricate the top bush and slide it down over the stanchion. To fit the bush into its recess it will be necessary to devise an alternative to the tubular drift tool used by Honda dealers. The best method is to use a length of tubing slightly bigger in diameter than the stanchion. Place a large plain washer against the bush and then tap it home using the tube as a form of slide-hammer. Take care not to scratch the stanchion during this

operation; it is best to ensure that it is pushed fully into the lower leg so that any accidental scoring is confined to the area above the seal.

11   Fit the special washer on top of the bush ensuring that its flat surface is facing downwards. Smear the lip of the new oil seal with ATF and slide it down the stanchion. Press the seal squarely into the lower leg as far as possible using hand pressure only and refit the plain washer. Rather than using the tubing to drive the oil seal fully into place (as in fitting the bush) it is preferable to use the old oil seal to press the new seal into place; this will ensure that the oil seal's top face is undamaged. Lever out the old oil seal afterwards. As soon as the circlip groove is exposed refit the seal retaining circlip, ensuring that its radiused edge is facing downwards and that it is correctly seated in its groove. Pack a small amount of grease above the oil seal and slide the dust seal into position.

12   Fill each leg with the specified amount and type of fork oil (see Specifications) and slowly pump the stanchion up and down to distribute the oil. Refit the fork spring, ensuring that it is installed with its closer-pitched coils uppermost, and refit the top bolt. Tighten the top bolt by hand only at this stage as it is much easier and safer to tighten it once the fork is securely clamped in the yokes.

10   Front forks: examination and renovation – CB650 SC (UK) and all US models

1   Wash all fork components thoroughly to remove all traces of old oil.

2   The parts most likely to wear are the bearing surfaces of the bushes. These control the play between the stanchion and lower leg and are designed to wear before damage occurs to the stanchion or lower leg. If there are any signs of scoring or obvious wear, the bushes must be renewed. Only in extreme cases will the stanchion or lower leg be worn; in these cases the affected item must also be renewed. Where access to the necessary measuring equipment can be gained, the actual degree of wear can be assessed by direct measurement. Any component that has worn beyond its service limit must be renewed.

3   The stanchions can be checked for straightness by rolling them on a flat surface such as a sheet of plate glass; any bending or distortion should immediately be evident. It is usually possible to straighten slightly bent stanchions, provided that the work is undertaken only by an expert; any local motorcycle dealer should be able to recommend such a person. However, if the stanchion is bent so much that the tubing has creased or even split, it must be renewed; straightening, even if possible, would induce severe stress, resulting in a fatigue failure at a later date.

4   Check that the stanchion surfaces are clean and free from chips, dents or corrosion which might weaken the tubing or cause oil seal failure. Use fine emery paper to polish off any corrosion; chips or dents, if minor, can be filled with Araldite or similar and rubbed down to restore the original shape when the filling compound has set. UK owners should note that such damage will cause the machine to fail its MOT test. If in doubt about the stanchion's strength, renew it in the interest of safety.

5   The fork springs will take a permanent set after considerable usage and will require renewal if the fork action becomes spongy. The degree of wear can be assessed by measuring their free length. If either spring has settled to less than the specified service limit, both springs should be renewed as a pair.

6   The oil seals should be renewed whenever they are disturbed, as should all sealing O-rings and washers. Check carefully the condition of each damper rod piston ring and renew it if there is any doubt about its condition. Also check the dust seals for signs of damage or deterioration, renewing them if necessary.

7   Make a careful check of all other fork leg components, checking for cracks in castings, damaged threads, a defective air valve and any other signs of wear or damage, renewing any faulty components.

11   Instrument heads: removal and replacement – CB650 SC models

1   Remove the two screws which retain the headlamp unit, disconnect the headlamp and parking lamp wiring (as applicable) and remove the unit from the machine. Unscrew the speedometer and tachometer cable retaining rings and disconnect the instrument panel wiring, situated inside the headlamp. Slacken and remove the instrument panel mounting nuts and lift the panel clear of its mounting bracket.

2   Remove the four screws which retain the instrument panel top cover and lift off the cover. Slacken the domed nuts which secure the covers to the bottom of each instrument, remove the covers and pull all the bulbholders out of the instrument head. Remove the nut which retains each instrument to the mounting bracket and separate the instrument heads and mounting bracket.

3   The instrument head itself is generally reliable, and is the least likely culprit in the event of a failure, this normally being attributed to the cable rather than the instrument mechanism. If however, it is noted that the speedometer has ceased to function whilst the odometer (mileage recorder) still functions, the instrument can be assumed to have failed. No form of home repair is practicable, and a replacement speedometer will be required. The only alternative is to seek the assistance of one of the companies who specialize in this kind of repair.

4   Refitting is a straightforward reversal of the removal sequence. Check that all mounting rubbers are in good condition, renewing them if necessary, and all bulbholders are correctly refitted. Spin the front wheel and turn the engine over to help engage the drive cables correctly with the instrument heads and tighten their retaining rings securely. Finally check that the instrument panel warning lamps and headlamp function correctly before taking the machine on the road.

12   Front wheel: removal and refitting – CB650 CB and SC models

1   Place the machine on its centre stand and raise the front wheel clear of the ground by placing a suitable block beneath the engine unit.

2   Expand the spring clip which secures the speedometer cable in position in its drive gearbox, and disconnect the cable. Slacken and remove the mounting bolts from either the left or right caliper assembly and slide the caliper off the disc. Tie the caliper to the frame to avoid placing any undue strain on the hydraulic hose. Before proceeding further, insert a wooden wedge between the brake pads; this will prevent the piston(s) being expelled should the brake lever be accidentally operated.

3   Slacken the front wheel spindle and withdraw the spindle whilst supporting the wheel. The wheel can then be lowered to the ground and manoeuvred out of the forks. Place another wooden wedge between the pads of the other caliper still mounted on the fork leg to prevent the piston(s) being accidentally expelled.

4   The front wheel is refitted by reversing the removal sequence. Before fitting the wheel, grease the oil seal lips and the speedometer gearbox. Refit the spacer to the right-hand side of the hub and fit the speedometer drive gearbox to the left-hand side ensuring that it locates correctly with the drive plate tangs. Remove the wooden wedges from both brake calipers.

5   Lift the wheel into position, ensuring that the brake disc is correctly positioned in the installed caliper and that the projection on the left-hand fork leg locates between the two lugs on the speedometer gearbox. Insert the wheel spindle from the right-hand side and tighten it to the specified torque setting.

6   Slide the brake caliper onto the disc and refit its mounting bolts, tightening them to the specified torque setting. Remove

the block from underneath the engine then apply the front brake and pump the front forks a few times to settle all components in position.

7   Using a 0.7 mm (0.028 in) feeler gauge, check the clearance between the left caliper's mounting bracket and the outer face of the disc. If the feeler gauge cannot be easily inserted pull the left lower leg outwards until the necessary clearance is obtained. Once the caliper clearance is correct, tighten the spindle pinch bolt to the specified torque setting.

8   Finally reconnect the speedometer cable and check the operation of the front brake before taking the machine on the road.

13   Front brake caliper: examination and renovation – CB650 C and SC models

Note: *Take great care not to spill hydraulic fluid on any painted or plastic cycle parts; it is a very effective paint stripper.*

1   When working on CB650 SC models dismantle the front brake calipers separately to avoid interchanging components.

2   To remove the caliper, slacken and remove the union bolt having first placed a suitable container underneath it in which to drain the fluid. Stop the flow of fluid from the reservoir by holding the front brake lever in against the handlebars; this is easily done using a stout elastic band.

3   When the fluid stops flowing from the hose union, clean the connections carefully and secure the hose end and fittings inside a clean polythene bag to await reassembly. It is most important to keep each component scrupulously clean, and to prevent the ingress of any foreign matter. For this reason, it is as well to prepare a clean area in which to work, before further dismantling. Ensure the outside of the caliper is thoroughly cleaned down.

4   Remove the brake pads as described in Section 3. On early CB650 C models remove the upper caliper body mounting bolt, and on later CB650 C and all SC models slide the caliper off of the mounting bracket guide pin. On all models remove the caliper from the machine.

5   Prise the anti-rattle pad spring out of the caliper, noting how

it is fitted, then prise off the two rubber boots from the mounting collar and slide the collar out of the caliper. Also remove the guide pin rubber boot from the caliper. The pistons may be expelled from the caliper body using an air jet (or a foot pump). Ensure that both pistons leave the bores at the same. If one sticks at any point the other piston must be restrained by firm hand pressure so that full pressure can overcome the resistance. It would be very difficult to extract one piston alone 'from this type of caliper without risking damage. *Under no circumstances should any attempt be made to lever or prise the pistons out of the caliper.* If the compressed air method fails, temporarily reconnect the flexible hose, refill the system, and use the handlebar lever pressure to displace the pistons hydraulically. Wrap some rag around the caliper to catch the inevitable shower of brake fluid. Whichever method is used take great care to avoid getting your fingers trapped by the emerging pistons. Once the pistons are out, remove both dust and fluid seals, taking great care not to damage the caliper bores.

6   Clean all components carefully, removing all traces of road dirt, friction material and corrosion. Note that only clean hydraulic fluid (or ethyl or isopropyl alcohol) should be used to clean hydraulic components; all normal cleaning solvents will attack the rubber seals. It is permissible to use a wire brush gently to remove dirt and corrosion except in the caliper bores and on the piston surfaces.

7   Renew both the fluid and dust seals as a matter of course. Never re-use an hydraulic seal after it has been disturbed and note that the piston seal must be in excellent condition as its secondary role is to return the piston when lever pressure is released, thus preventing brake drag. Carefully examine the mounting collar rubber boots, renewing them if they are perished split or otherwise damaged. Discard the sealing washers which are fitted each side of the brake hose union; these must be renewed as a matter of course.

8   Examine the piston surface and caliper bore for signs of wear and damage, normally caused by the presence of road dirt or corrosion. If wear or deep scoring or scratches are found, which might cause fluid leakage, the component concerned must be renewed. If access to the necessary measuring equipment is available, the degree of wear can be assessed by direct measurement. If any component has worn to beyond its service limit, given in the Specifications at the start of this Chapter, it must be renewed.

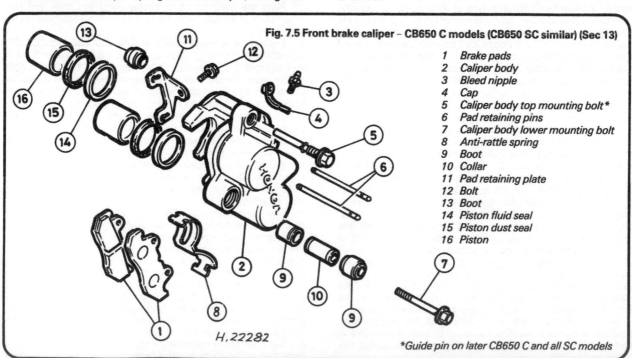

Fig. 7.5 Front brake caliper – CB650 C models (CB650 SC similar) (Sec 13)

1   *Brake pads*
2   *Caliper body*
3   *Bleed nipple*
4   *Cap*
5   *Caliper body top mounting bolt* *
6   *Pad retaining pins*
7   *Caliper body lower mounting bolt*
8   *Anti-rattle spring*
9   *Boot*
10 *Collar*
11 *Pad retaining plate*
12 *Bolt*
13 *Boot*
14 *Piston fluid seal*
15 *Piston dust seal*
16 *Piston*

H.22282

*Guide pin on later CB650 C and all SC models

RECTIFIER   UNIT : kΩ

(−) Probe \ Probe (+)	Red/White	Green	Yellow 1	Yellow 2	Yellow 3
Red/White		∞	∞	∞	∞
Green	0.5 – 50		0.5 – 50	0.5 – 50	0.5 – 50
Yellow 1	0.5 – 50	∞		∞	∞
Yellow 2	0.5 – 50	∞	∞		∞
Yellow 3	0.5 – 50	∞	∞	∞	

REGULATOR   UNIT : kΩ

(−) Probe \ Probe (+)	Black	White	Green
Black		1 – 30	0.5 – 20
White	0.5 – 30		1 – 50
Green	0.5 – 20	0.5 – 30	

H.16624

Fig. 7.6 Regulator/rectifier test table – CB650 SC (UK) and all US models (Sec 14)

9   Inspect the caliper mounting collar and caliper bracket guide pin or bolt (as appropriate), and their respective bearing surfaces, for signs of wear or damage such as scoring or corrosion. Light traces of corrosion may be removed with fine emery paper, but if the surface has become pitted or scored the damaged component must be renewed.

10   On reassembly, soak the new fluid and dust seals in clean hydraulic fluid and fit them into the caliper bores, ensuring that each is correctly seated in its groove. Smear hydraulic fluid over the caliper bore and piston surface and refit the pistons, rotating them slightly whilst keeping them square to the caliper bore so that they do not stick or displace the seals.

11   Lubricate the caliper guide pin/mounting bolts (as applica-

ble) and mounting collar with silicone grease and fit the collar and three rubber boots to the caliper, ensuring that each boot is correctly seated. Refit the anti-rattle pad spring to the caliper. Locate the caliper with its mounting bracket guide pin or refit the upper mounting bolt (as applicable) and ensure that the rubber boot locates correctly with the caliper mounting bracket.

12   Refit the brake pads as described in Section 3. Ensure the caliper body to bracket bolt(s) are tightened to the specified torque setting and check that the caliper slides smoothly and easily on its mounting bracket. Position a new sealing washer on each side of the hose union and tighten the bolt to the specified torque setting. Bleed the system as described in Chapter 5, after filling the reservoir with new hydraulic fluid, then check for fluid leakage whilst applying the brake. Push the machine forwards and bring it to a halt by applying the brake. Do this several times to ensure that the brake is working correctly before taking the machine for a test run. During the run, use the brake as often as possible and on completion, recheck for signs of fluid loss.

14   Regulator/rectifier: testing – CB650 SC (UK) and all US models

1   The regulator/rectifier unit fitted to these models is similar to that fitted on earlier models but can be tested as follows.

2   Remove the left-hand side panel and disconnect both block connectors from the regulator/rectifier. Using a multimeter set to the appropriate scale, measure the resistances across the various block connector terminals on the regulator/rectifier side of the wiring. If the readings obtained differ greatly from those shown in the accompanying table it is likely that the regulator/rectifier unit is faulty and should be renewed. However, before consigning the unit to the scrap bin, it is recommended that the unit be taken to an authorised Honda Service Agent for further testing to confirm your test results. Note that the readings given in the table are for a positive (+) earth meter, the exact type recommended by Honda being either the Sanwa Electrical Tester (Pt No 07308-0020000) set to the K ohm range, or the Kowa Tester (Model TH-SH) set to the x 100 ohms range. If the meter being used has a negative (–) earth, reverse the meter probes to obtain the correct readings.

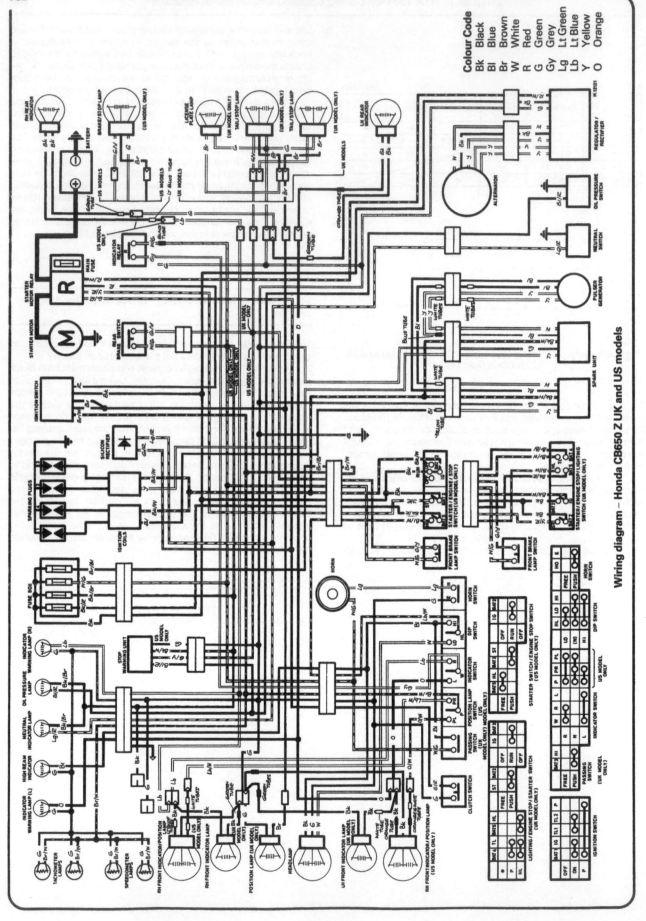

Wiring diagram – Honda CB650 Z UK and US models

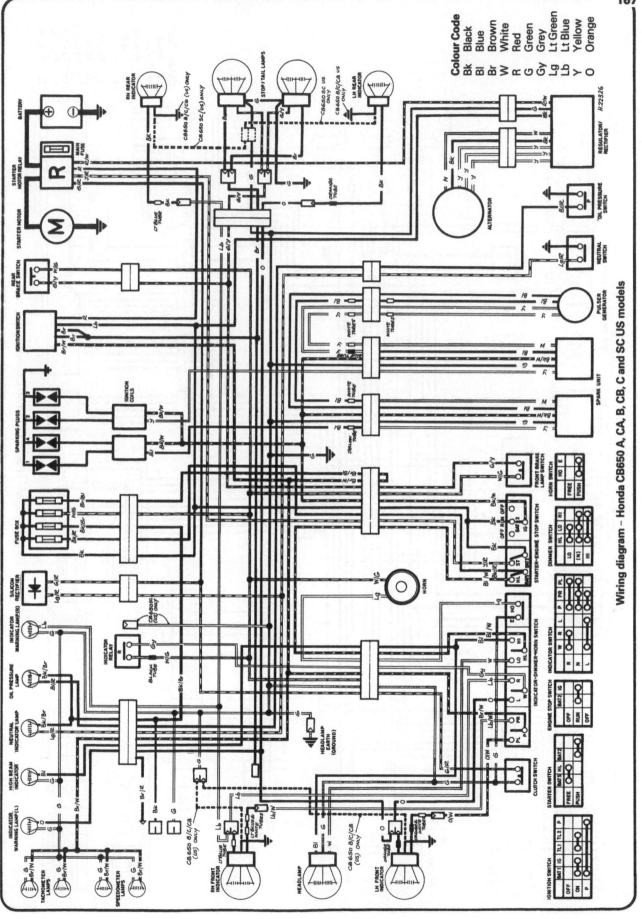

Wiring diagram – Honda CB650 A, CA, B, CB, C and SC US models

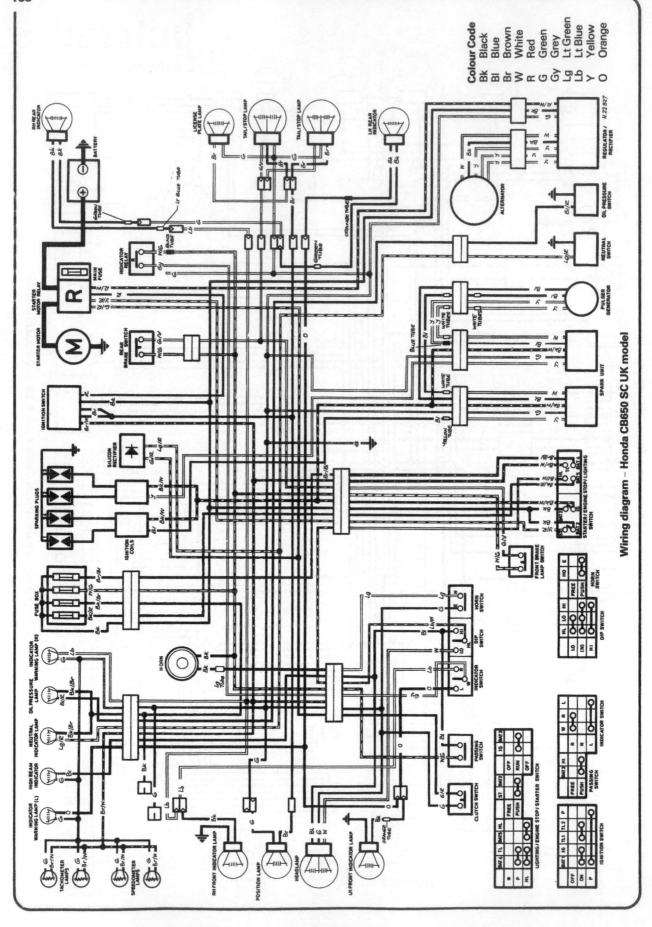

Wiring diagram – Honda CB650 SC UK model

English/American terminology

Because this book has been written in England, British English component names, phrases and spellings have been used throughout. American English usage is quite often different and whereas normally no confusion should occur, a list of equivalent terminology is given below.

English	American	English	American
Air filter	Air cleaner	Number plate	License plate
Alignment (headlamp)	Aim	Output or layshaft	Countershaft
Allen screw/key	Socket screw/wrench	Panniers	Side cases
Anticlockwise	Counterclockwise	Paraffin	Kerosene
Bottom/top gear	Low/high gear	Petrol	Gasoline
Bottom/top yoke	Bottom/top triple clamp	Petrol/fuel tank	Gas tank
Bush	Bushing	Pinking	Pinging
Carburettor	Carburetor	Rear suspension unit	Rear shock absorber
Catch	Latch	Rocker cover	Valve cover
Circlip	Snap ring	Selector	Shifter
Clutch drum	Clutch housing	Self-locking pliers	Vise-grips
Dip switch	Dimmer switch	Side or parking lamp	Parking or auxiliary light
Disulphide	Disulfide	Side or prop stand	Kick stand
Dynamo	DC generator	Silencer	Muffler
Earth	Ground	Spanner	Wrench
End float	End play	Split pin	Cotter pin
Engineer's blue	Machinist's dye	Stanchion	Tube
Exhaust pipe	Header	Sulphuric	Sulfuric
Fault diagnosis	Trouble shooting	Sump	Oil pan
Float chamber	Float bowl	Swinging arm	Swingarm
Footrest	Footpeg	Tab washer	Lock washer
Fuel/petrol tap	Petcock	Top box	Trunk
Gaiter	Boot	Torch	Flashlight
Gearbox	Transmission	Two/four stroke	Two/four cycle
Gearchange	Shift	Tyre	Tire
Gudgeon pin	Wrist/piston pin	Valve collar	Valve retainer
Indicator	Turn signal	Valve collets	Valve cotters
Inlet	Intake	Vice	Vise
Input shaft or mainshaft	Mainshaft	Wheel spindle	Axle
Kickstart	Kickstarter	White spirit	Stoddard solvent
Lower leg	Slider	Windscreen	Windshield
Mudguard	Fender		

Index